Plant Physiology: From Omic Analysis toward Physiological Mechanism Research

Plant Physiology: From Omic Analysis toward Physiological Mechanism Research

Editors

Jie Luo
Sen Meng

MDPI

Basel • Beijing • Wuhan • Barcelona • Belgrade • Novi Sad • Cluj • Manchester

Editors
Jie Luo
College of Horticulture & Forestry Sciences
Huazhong Agricultural University
Wuhan
China

Sen Meng
State Key Laboratory of Tree Genetics and Breeding
Chinese Academy of Forestry
Guangzhou
China

Editorial Office
MDPI
St. Alban-Anlage 66
4052 Basel, Switzerland

This is a reprint of articles from the Special Issue published online in the open access journal *Life* (ISSN 2075-1729) (available at: https://www.mdpi.com/journal/life/special_issues/Omic_Plant_Physiology).

For citation purposes, cite each article independently as indicated on the article page online and as indicated below:

Lastname, A.A.; Lastname, B.B. Article Title. *Journal Name* **Year**, *Volume Number*, Page Range.

ISBN 978-3-7258-1438-1 (Hbk)
ISBN 978-3-7258-1437-4 (PDF)
doi.org/10.3390/books978-3-7258-1437-4

Contents

life

MDPI

Editorial

From Omics Analysis toward Physiological Mechanism Research in Plants

Yan Lu [1], Sen Meng [2,*] and Jie Luo [3,*]

[1] Jiangsu Key Laboratory for the Research and Utilization of Plant Resources, Institute of Botany, Jiangsu Province and Chinese Academy of Sciences (Nanjing Botanical Garden Memorial Sun Yat-Sen), Nanjing 210014, China; 18260093866@163.com

[2] State Key Laboratory of Tree Genetics and Breeding, Research Institute of Tropical Forestry, Chinese Academy of Forestry, Guangzhou 510520, China

[3] College of Horticulture and Forestry Sciences, Hubei Engineering Technology Research Center for Forestry Information, Huazhong Agricultural University, Wuhan 430070, China

* Correspondence: mengsen021124@126.com (S.M.); luojie@mail.hzau.edu.cn or jie.luo@vip.163.com (J.L.)

With the development of big data in system biology researches, the high-throughput omics analysis has become the most popular high technology in the fields of plant research [1,2]. Increasing studies have applied multi-omics such as genomics, transcriptomics, proteomics and metabolome to uncover the underlying mechanisms of plant physiology, bringing many new discoveries [3–6]. Undoubtedly, it is significant to combine omics and traditional physiology techniques to explore the insights into plant development, abiotic stresses responses, hormonal and environmental signaling and key trait formation. In this Special Issue, a total of ten papers were collected, including eight research papers and two review papers.

Cutting propagation is an important vegetative propagation method for rapid and large-scale reproduction samplings in agricultural practices [7]. Many plants can form adventitious root easily with cuttings without any treatment, but some plants need special treatments for efficient rooting [8]. Exogenous applications of hormones such as auxins are widely used for stimulating cutting rooting capacities, especially for some difficult-to-root plants [8–10]. Quan et al. compared different effects of phytohormones on the rooting capacities of *Catalpa bignonioides* softwood cuttings and found that applications of 1 g L^{-1} IBA significantly promoted rooting rate compared to control and other growth regulator treatments (Contribution 1). The exogenous application of growth regulators also triggered the changes in endogenous hormone levels and ratios in the phloem of *C. bignonioides* softwood cuttings, which might be beneficial for rooting (Contribution 1). In another paper, Sun et al. evaluated the effects of different types and concentrations of hormones as well as the influence of soaking times on the rooting capacities of *Morus* 'Yueshenda 10' softwood cuttings and found that applications of 800 mg L^{-1} commercial rooting powder ABT1 for 30 min could significantly improve the rooting ratio up to 66.24% (Contribution 2). These two papers provided a reference to select optimal hormones for stimulating the rooting ratio of softwood cuttings in woody plants.

Santalum album L. is a semi-parasitic evergreen tree with important economic value [11]. Taking advantage of the sandalwood transcriptome, Zhang et al. identified eight basic helix–loop–helix (*bHLH*) gene member using bioinformation methods (ontribution 3). The authors further found that SaMYC1 could function as transcription factors and activate the expression of two genes (*SaSSy* and *SaCYP736A167*) that participate in the biosynthesis of the main sandal sesquiterpenes (Contribution 3). Similarly, Chen et al. identified nine plant-specific Rac/Rop small GTPases gene members with the full-length transcriptome data of *S. album* and evaluated the gene expression levels of *Rac* genes in response to drought and hormone treatments (Contribution 4). The authors also identified several key *Rac* genes that closely related to the formation of haustoria in *S. album*, which is

Citation: Lu, Y.; Meng, S.; Luo, J. From Omics Analysis toward Physiological Mechanism Research in Plants. *Life* **2023**, *13*, 2275. https://doi.org/10.3390/life13122275

Received: 24 November 2023
Accepted: 28 November 2023
Published: 29 November 2023

important for their semi-parasitic lifestyle (Contribution 4). Due to the lack of complete genome information and effective transgenic methods for *S. album*, a gene functional study could not be applied in *S. album*, which might be a bottleneck to understand the molecular mechanism underlying the growth and environmental adaptations in *S. album* for now. However, with the acknowledgement obtained from the bioinformation analysis combined with molecular techniques, the fundamental molecular mechanisms underlying the synthesis of sandal sesquiterpenes and formation of haustoria will be uncovered soon in the future.

Volatile aroma is an important trait determining the quality of fruits [12]. Taking advantage of RNA-seq, Li et al. conducted a time-series transcriptomic analysis in nine sample stages from fruit development to the fruit storage to uncover the molecular network underlying the volatile aroma formation in pears (Contribution 5). Several key modules related to fatty acid formation were detected using the WGCNA methods, and hub genes in the modules related to the volatile aroma were identified by gene co-expression analysis (Contribution 5).

Disrupted meiotic cDNA (*DMC1*) is an important gene that controls DNA recombination through crossing over in meiosis [13]. Kumar et al. applied the RNA interference approach to produce asynaptic mutants by knockdown of the *StDMC1* gene (Contribution 6). The *StDMC1* RNAi lines significantly reduced the pollen viability compared to Kufri Jyoti control plants, with reduced expression levels of the *StDMC1* gene (Contribution 6). *Isodon rubescens* (Hemsl.) Hara is an important medicinal plant in China; its active ingredients show anti-tumor effects [14]. Lian et al. cloned a UDP-glycosyltransferase gene (*UGT*) *IrUGT86A1-like* in *I. rubescens* and predicted the protein properties and cellular location (Contribution 7). The *IrUGT86A1-like* gene was highly expressed in leaves and its expression levels in tissues were highly correlated to the oridonin contents (Contribution 7). Furthermore, gene expression analysis revealed that *IrUGT86A1-like* was positively regulated by NaCl and MeJA treatments, but negatively regulated by abscisic acid (ABA) treatment (Contribution 7). Chen et al. reported that the maize *Constitutively Photomorphogenic 1* (*ZmCOP1*) gene could significantly promote maize mesocotyl lengths and plant height (Contribution 8). By using RNA-seq techniques, the authors identified several different expression genes between B73 and the *zmcop1-1* mutant; pathway analysis indicated the major changes in phytohormone signal transduction (Contribution 8). The results collectively supported that the *ZmCOP1* gene regulates maize architecture by affecting hormone pathways, the underlying mechanisms need to be further explored.

Two review papers were published in this Special Issue. Lin et al. nicely summarized the recent advances of phosphate-binding loop guanosine triphosphatases (P-loop GTPases) in plant growth and stress response (Contribution 9). The authors introduced in detail the structure of the G domain of G proteins, the G domain of YchF, and compared different protein structures (Contribution 9). The critical roles played by YchF in plant oxidative stress, environmental stress response, protein biosynthesis and degradation, and maintaining proteostasis were extensively reviewed by the authors (Contribution 9). miRNAs play crucial roles in plant growth and development, as well as environmental responses. Teng et al. reviewed the recent advances in the transcriptional regulation of miRNAs in plants (Contribution 10). The authors summarized the recent methods used for identifying microRNA promoters, and the important roles played by miRNA in plant growth and development, synthesis of secondary metabolites, disease resistance, abiotic stress, phytohormone signaling pathways, and so on (Contribution 10).

These research papers collectively highlight the significance of applying multiple techniques to understand the mechanisms underlying the physiology of plant development, environmental stresses, and key trait formation. Key genes involved in these physiological processes have been identified, and it is significant for us to build targeted strategies for plant resource protection and utilization. Moreover, these studies encompass a wide array of plant species, mainly including non-timber woody plants, medical plants and crops.

As guest editors, we believe that this Special Issue offers a valuable compilation of research findings that contribute to omics analysis toward physiological mechanism in plants. The diversified species and omics technique outline the potential of combining omics and physiological analyses to enhance plant conservation, sustainable production and utilization. However, more research using innovative omics techniques such as ATAC-seq and single cell RNA-seq to unravel the underlying mechanisms of plant physiology is urgently needed to further expand our knowledge. We thank all contributing authors for their work on this Special Issue.

Author Contributions: Y.L., S.M. and J.L. wrote the paper. All authors have read and agreed to the published version of the manuscript.

Funding: The study was supported by the Fundamental Research Funds for the Central Universities (Grant no. 2262022YLYJ007).

Conflicts of Interest: The authors declare no conflict of interest.

List of Contributions

1. Quan, J.e.; Ni, R.; Wang, Y.; Sun, J.; Ma, M.; Bi, H. Effects of different growth regulators on the rooting of *Catalpa bignonioides* softwood cuttings. *Life* **2022**, *12*, 1231.
2. Sun, J.; Li, H.; Chen, H.; Wang, T.; Quan, J.e.; Bi, H. The effect of hormone types, concentrations, and treatment times on the rooting traits of *Morus* 'Yueshenda 10' softwood cuttings. *Life* **2023**, *13*, 1032.
3. Zhang, T.; Chen, X.; Xiong, Y.; Niu, M.; Zhang, Y.; Yan, H.; Li, Y.; Zhang, X.; Ma, G. Identification and functional analysis of *SabHLHs* in *Santalum album* L. *Life* **2022**, *12*, 1017.
4. Chen, Y.; Wang, S.; Liu, X.; Wang, D.; Liu, Y.; Hu, L.; Meng, S. Analysis of Rac/Rop Small GTPase family expression in *Santalum album* L. and their potential roles in drought stress and hormone treatments. *Life* **2022**, *12*, 1980.
5. Li, H.; Quan, J.; Rana, S.; Wang, Y.; Li, Z.; Cai, Q.; Ma, S.; Geng, X.; Liu, Z. The molecular network behind volatile aroma formation in pear (*Pyrus* spp. Panguxiang) revealed by transcriptome profiling via fatty acid metabolic pathways. *Life* **2022**, *12*, 1494.
6. Kumar, A.; Siddappa, S.; Bhardwaj, V.; Dalamu; Singh, B.; Sharma, N.; Dipta, B.; Kumar, V.; Goutam, U.; Sood, S. Generation of asynaptic mutants in potato by disrupting *StDMC1* gene using RNA interference approach. *Life* **2023**, *13*, 174.
7. Lian, C.; Lan, J.; Zhang, B.; Yang, H.; Guo, K.; Li, J.; Chen, S. Molecular cloning and functional analysis of *IrUGT86A1-like* gene in medicinal plant *Isodon rubescens* (Hemsl.) Hara. *Life* **2022**, *12*, 1334.
8. Chen, L.; Li, Q.; Wang, M.; Xiao, F.; Li, K.; Yang, R.; Sun, M.; Zhang, H.; Guo, J.; Chen, J.; et al. *ZmCOP1* regulates maize mesocotyl length and plant height through the phytohormone pathways. *Life* **2023**, *13*, 1522.
9. Lin, Z.; Li, R.; Han, Z.; Liu, Y.; Gao, L.; Huang, S.; Miao, Y.; Miao, R. The universally conserved unconventional G protein YchF is critical for growth and stress response. *Life* **2023**, *13*, 1058.
10. Teng, C.; Zhang, C.; Guo, F.; Song, L.; Fang, Y. Advances in the study of the transcriptional regulation mechanism of plant miRNAs. *Life* **2023**, *13*, 1917.

References

1. Mochida, K.; Shinozaki, K. Advances in omics and bioinformatics tools for systems analyses of plant functions. *Plant Cell Physiol.* **2011**, *52*, 2017–2038. [CrossRef]
2. Liang, S.; Li, Y.; Chen, Y.; Huang, H.; Zhou, R.; Ma, T. Application and prospects of single-cell and spatial omics technologies in woody plants. *For. Res.* **2023**, *3*, 27. [CrossRef]
3. Luo, J.; Havé, M.; Clément, G.; Tellier, F.; Balliau, T.; Launay-Avon, A.; Guérard, F.; Zivy, M.; Masclaux-Daubresse, C. Integrating multiple omics to identify common and specific molecular changes occurring in Arabidopsis under chronic nitrate and sulfate limitations. *J. Exp. Bot.* **2020**, *71*, 6471–6490. [CrossRef]

4. Voelckel, C.; Gruenheit, N.; Lockhart, P. Evolutionary transcriptomics and proteomics: Insight into plant adaptation. *Trends Plant Sci.* **2017**, *22*, 462–471. [CrossRef]
5. Naik, B.; Kumar, V.; Rizwanuddin, S.; Chauhan, M.; Choudhary, M.; Gupta, A.K.; Kumar, P.; Kumar, V.; Saris, P.E.J.; Rather, M.A. Genomics, proteomics, and metabolomics approaches to improve abiotic stress tolerance in tomato plant. *Int. J. Mol. Sci.* **2023**, *24*, 3025. [CrossRef] [PubMed]
6. Cai, K.; Zhao, Q.; Li, H.; Zhang, Q.; Li, Y.; Han, R.; Jiang, T.; Pei, X.; Zhang, L.; Zhao, X. Deciphering aroma formation during flowering in nectar tree (*Tilia amurensis*): Insights from integrated metabolome and transcriptome analysis. *For. Res.* **2023**, *3*, 24. [CrossRef]
7. Ren, Z.-M.; Zhang, D.; Jiao, C.; Li, D.-Q.; Wu, Y.; Wang, X.-Y.; Gao, C.; Lin, Y.-F.; Ruan, Y.-L.; Xia, Y.-P. Comparative transcriptome and metabolome analyses identified the mode of sucrose degradation as a metabolic marker for early vegetative propagation in bulbs of *Lycoris*. *Plant J.* **2022**, *112*, 115–134. [CrossRef] [PubMed]
8. Shang, C.; Yang, H.; Ma, S.; Shen, Q.; Liu, L.; Hou, C.; Cao, X.; Cheng, J. Physiological and transcriptomic changes during the early phases of adventitious root formation in mulberry stem hardwood cuttings. *Int. J. Mol. Sci.* **2019**, *20*, 3707. [CrossRef] [PubMed]
9. Quan, J.; Zhang, C.; Zhang, S.; Meng, S.; Zhao, Z.; Xu, X. Molecular cloning and expression analysis of the *MTN* gene during adventitious root development in IBA-induced tetraploid black locust. *Gene* **2014**, *553*, 140–150. [CrossRef] [PubMed]
10. Quan, J.; Meng, S.; Guo, E.; Zhang, S.; Zhao, Z.; Yang, X. De novo sequencing and comparative transcriptome analysis of adventitious root development induced by exogenous indole-3-butyric acid in cuttings of tetraploid black locust. *BMC Genom.* **2017**, *18*, 179. [CrossRef] [PubMed]
11. Chen, X.; Zhang, Y.; Yan, H.; Niu, M.; Xiong, Y.; Zhang, X.; Li, Y.; da Silva, J.A.T.; Ma, G. Cloning and functional analysis of 1-deoxy-d-xylulose-5-phosphate synthase (DXS) in *Santalum album* L. *Gene* **2023**, *851*, 146762. [CrossRef] [PubMed]
12. El Hadi, M.A.M.; Zhang, F.-J.; Wu, F.-F.; Zhou, C.-H.; Tao, J. Advances in fruit aroma volatile research. *Molecules* **2013**, *18*, 8200–8229. [CrossRef] [PubMed]
13. Klimyuk, V.I.; Jones, J.D. *AtDMC1*, the Arabidopsis homologue of the yeast *DMC1* gene: Characterization, transposon-induced allelic variation and meiosis-associated expression. *Plant J.* **1997**, *11*, 1–14. [CrossRef] [PubMed]
14. Chen, X.; Dai, X.; Liu, Y.; He, X.; Gong, G. *Isodon rubescens* (Hemls.) Hara.: A comprehensive review on traditional uses, phytochemistry, and pharmacological activities. *Front. Pharmacol.* **2022**, *13*, 766581. [CrossRef] [PubMed]

life

MDPI

Review

Advances in the Study of the Transcriptional Regulation Mechanism of Plant miRNAs

Caixia Teng †, Chunting Zhang †, Fei Guo, Linhong Song and Yanni Fang *

College of Horticulture and Forestry Science, Huazhong Agricultural University, Wuhan 430070, China; caixiateng2000@163.com (C.T.); 17684570598@163.com (C.Z.); guofei@mail.hzau.edu.cn (F.G.)
* Correspondence: fangyanni@mail.hzau.edu.cn
† These authors contributed equally to this work.

Abstract: MicroRNAs (miRNA) are a class of endogenous, non-coding, small RNAs with about 22 nucleotides (nt), that are widespread in plants and are involved in various biological processes, such as development, flowering phase transition, hormone signal transduction, and stress response. The transcriptional regulation of miRNAs is an important process of miRNA gene regulation, and it is essential for miRNA biosynthesis and function. Like mRNAs, miRNAs are transcribed by RNA polymerase II, and these transcription processes are regulated by various transcription factors and other proteins. Consequently, the upstream genes regulating miRNA transcription, their specific expression, and the regulating mechanism were reviewed to provide more information for further research on the miRNA regulatory mechanism and help to further understand the regulatory networks of plant miRNAs.

Keywords: plant; miRNAs; promoter; transcription regulation; molecular mechanism

Citation: Teng, C.; Zhang, C.; Guo, F.; Song, L.; Fang, Y. Advances in the Study of the Transcriptional Regulation Mechanism of Plant miRNAs. *Life* **2023**, *13*, 1917. https://doi.org/10.3390/life13091917

Academic Editor: Kousuke Hanada

Received: 11 August 2023
Revised: 12 September 2023
Accepted: 13 September 2023
Published: 15 September 2023

1. Introduction

Small RNAs, which range in size from 20 to 24 nucleotides, are derived from dsRNAs through processing mediated by the RNase III enzyme [1]. They can be categorized into several groups based on differences in their biogenesis and function [1]. miRNAs and small interfering RNAs (siRNAs) are two major classes of endogenous small RNAs in plants. siRNAs can be subdivided into *trans*-acting siRNAs, repeat-associated siRNAs (rasiRNAs), heterochromatic siRNAs (hc-siRNAs), and nat-siRNAs in plants (nat-siRNAs) [1,2].

Most plant miRNAs originate from intergenic regions and are transcribed from their own promoters [3]. Some other miRNAs originate from non-coding RNAs or the introns of coding genes [3]. Under the action of RNA polymerase II (Pol II), miRNA genes are transcribed into precursor transcript pri-miRNAs in the nucleus. With the help of Dicer enzymes DCL1, pre-miRNAs are cleaved from pri-miRNAs, form into stem–loop structures, and are subsequently cleaved into double-stranded miRNA/miRNA* complexes [3,4]. After the miRNA* is degraded, the mature miRNA binds to the Argonaute1 (AGO1) protein, which forms the RNA-induced silencing complex (RISC) to regulate the expressions of target genes post-transcriptionally through direct transcript cleavage or translation repression [5,6]. To analyze the biological functions and regulatory mechanism mediated by miRNAs, many target genes have been identified using degradome sequencing, and some of them have been validated in vitro or in vivo in recent years. Most target genes of conserved miRNA are transcription factors, such as MYB (myeloblastosis oncogene), SPL (AQUAMOSA promoter-binding protein-like), NAC (NAM, ATAF1/2, CUC1/2), AP2 (APETALA2), the Zinc finger protein HD-ZIP (homeodomain leucine zipper) family, GRF (growth-regulating factor), and ARF (auxin-responsive factor), etc. (Figure 1) [6–10], which make miRNAs key players in the plant regulatory network [11,12]. In addition to transcription factors, some miRNAs target resistance genes [13], ubiquitin-conjugating enzymes, and other genes [14]. Numerous studies have shown that miRNAs play wide and important roles in plant growth and development (Figure 1), the flowering cycle, hormone signal transduction, the stress response, and so on [8,15–23].

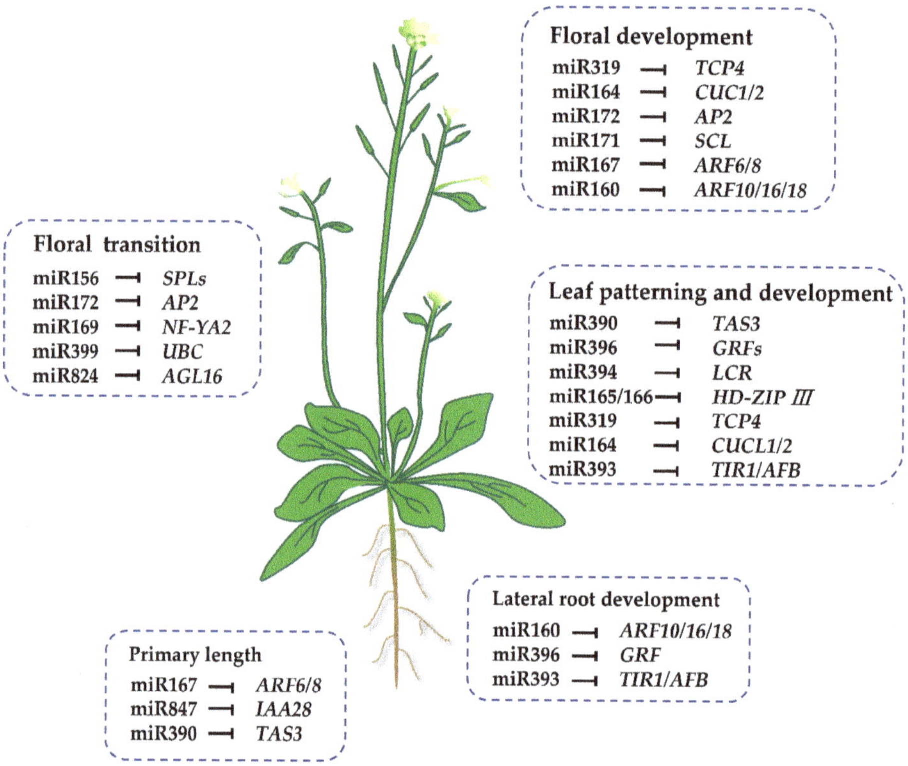

Figure 1. Regulatory functions of miRNAs on plant growth and development. *MYB* (*myeloblastosis oncogene*), *SPL* (*AQUAMOSA promoter-binding protein-like*), *NAC* (*NAM, ATAF1/2, CUC1/2*), *AP2* (*APETALA2*), *GRF* (*growth-regulating factor*), *ARF* (*auxin-responsive factor*), *CUC* (*cup-shaped cotyledon*), *AGL* (*AGAMOUS-Like*), *HD-ZIP III* (*class III homeodomain leucine zipper*), *UBC* (*ubiquitin-conjugating enzyme*), *TAS* (*trans-acting short-interfering RNA*), *LCR* (*leaf curling responsiveness*), *TCP* (*teosinte branched*), *NF-YA2* (*nuclear transcription factor Y subunit alpha*), *SCL* (*SCARECROW-Like*), *and AFB* (*auxin receptor F box protein*).

Transcriptional regulation is important for miRNA expression. Like protein-coding genes, the transcription of miRNA genes is regulated by various transcription factors and other proteins. In general, transcription factors regulate the specific expressions of miRNA genes through binding to the DNA-binding domains and *cis*-acting elements on the miRNA promoters. In addition to transcription factors, some proteins can regulate miRNA transcription by binding to RNA polymerase II to affect the accumulation of polymerase in the miRNA promoter region. It is widely known that miRNAs play significant roles in various processes of plant development by targeting and regulating many genes post-transcriptionally. But how are these miRNAs transcribed from miRNA genes? What studies have been conducted in plants on the upstream regulators of miRNAs? What proteins or transcription factors regulate the transcription of miRNAs to obtain spatiotemporal-specific expression, and how do miRNAs obtain transcription and expression in response to external signals to adapt to environmental changes? To answer these questions, studies on the upstream regulatory factors of miRNAs have been carried out in several plants to analyze the spatiotemporal-specific expressions and transcriptional regulatory mechanisms of miR-

NAs. Therefore, this review summarizes the progress of the research on the transcriptional regulatory factors of miRNA genes, which will help to deepen our understanding of the regulatory network and molecular mechanisms of miRNAs in the plant lifecycle.

2. Identification of Promoters and *cis*-Acting Elements of miRNAs

Many plant miRNA genes are located in intergenic regions as independent transcriptional units. A few miRNAs are located in the intron region of protein-coding genes, which is co-transcribed with the host genes [3,24]. Like coding genes, eukaryotic miRNA genes are transcribed by Pol II under the regulation of general and specific transcription factors [25,26]. The miRNA promoter consists of the core promoter region and distal upstream region. The core promoter region contains elements such as an initiator, TATA box motifs, CAAT box motifs, and *cis*-acting elements. The initiator is a conserved sequence located near the Transcription Start Site (TSS), while the TATA box is a conserved AT-rich sequence about 30 bases upstream of the TSS, which regulates transcription initiation together with the initiator. The CAAT box is located about 85 bases upstream of the TSS and controls the frequency of transcription initiation. The upstream distal region comprises multiple *cis*-acting elements that specifically bind to *trans*-acting factors to synergistically regulate the transcription of miRNAs and the spatiotemporal-specific expressions of miRNAs. *Trans*-acting siRNAs (ta-siRNA) are a class of endogenous small RNAs that are produced from non-coding *TAS* genes. nat-siRNAs are derived from the overlapping transcript of two adjacent genes located on opposite strands [27]. *TAS* genes are transcribed from their own promoter by the RNA polymerase II as long primary RNAs [28]. nat-siRNA biogenesis also relies on the transcription of a pair of antisense genes produced by RNA polymerase II. Therefore, the precursors of ta-siRNAs and nat-siRNAs are transcribed by the Pol II promoter, which is similar to miRNA. Different from the other classes of small RNAs, the precursors of hc-siRNAs are generated by RNA polymerase IV on repetitive regions and transposable elements [29].

miRNA promoter and *cis*-acting element identification is important for transcriptional regulation analyses of miRNAs. Previously, researchers identified 63 miRNA transcriptional start sites in *Arabidopsis* using 5'RACE technology and found that the majority of miRNA promoters contained TATA-boxes [30]. In recent years, the promoters of many miRNA genes have been identified in a variety of plants using multiple bioinformatic prediction methods (Table 1). By applying their self-developed computational sequence-centric method, common query voting (CoVote), Zhou et al. [26] predicted the putative core promoters for the most known intergenic miRNA genes of *Arabidopsis* and rice. TSSP (http://linux1.softberry.com, accessed on 12 September 2023) is a tool that predicts the TSS, combing characteristics describing the functional motifs of common core promoters and the composition of these sites. The promoters of miRNAs can be obtained after TSSs are predicted based on the general rule that the promoter region of each gene is located 1500 bp upstream of the TSS [31]. Using the promoter prediction method developed by Zhou et al. [26], the sequences for the TSS were predicted through the TSSP database after searching the 2000 bp upstream of the 5′ end of the pre-miRNA or the sequences between 400 bp downstream of the neighboring protein-coding gene and the pre-miRNA. By these means, a total of 249 promoter sequences for 158 miRNAs precursors in rice [32], 229 promoters for 139 miRNA precursors in poplar [33], 191 promoters for 122 miRNA loci downloaded in the miRBase, and 64 TSSs for 22 phosphorus-deficient responsive miRNAs in soybean [34,35] have been successfully identified. Additionally, 132 TSSs of 42 miRNA were discovered in *Arabidopsis* using a computational method developed from the genome-wide profiles of nine histone markers, including H3K4me2, H3K4me3, H3K9Ac, H3K9me2, H3K18Ac, H3K27me1, H3K27me3, H3K36me2, and H3K36me3 [36]. A total of 699 promoters and 440 miRNA TSSs have been predicted in soybean using degradome libraries and the TSSP software [37]. In total, 21 high-quality promoters of 23 intergenic miRNAs in cassava were predicted via a hybrid computational method combining PromPredict and the TSSP software based on the DNA free energy change and a common core regulatory element analysis [38]. Another computational sequence-centric method, named the Query-Ranked Frequent Rule (QRFR), was developed by Zhou et al. for identifying the core promoter regions of miRNA genes [39]. In total, 47 core promoters of 40 miRNA genes in *Arabidopsis* studied in [30] were tested using the QRFR, and 34 were correctly confirmed [30,39].

Table 1. Genomic identification methods of microRNA promoters.

Species	Counts of miRNAs Loci	Counts of Identified miRNAs Promoters	Identification Methods	Reference
Arabidopsis	52	63	5′ RACE	[30]
Arabidopsis	95	98	Common query voting (CoVote)	[26]
Rice	114	104	CoVote	[26]
Rice	158	249	TSSP	[32]
Soybean	22	64	TSSP	[35]
Soybean	12	191	TSSP	[34]
Populus	139	229	TSSP	[33]
Soybean	440	699	Degradome libraries and TSSP	[37]
Soybean	298	132	Genome-wide profiles of nine histone markers	[36]
Arabidopsis	40	34	Query-Ranked Frequent Rule (QRFR)	[39]
Cassava	23	21	PromPredict and TSSP	[38]

Today, small RNA sequencing and computational prediction technology have been rapidly developed to help with the genome-wide identification of miRNAs and their precursors. Unlike for protein-coding genes, the distance from TSSs to pre-miRNAs is longer and more irregular. Presently, except for model plants such as *Arabidopsis*, the TSSs of the miRNAs in other higher plants are mainly obtained through biotech software prediction due to the high cost of experimental methods. Consequently, the promoter identification of miRNAs in previous publications has usually been limited to beginning upstream from mature or pre-miRNAs due to the lack of the exact TSS information of the miRNAs. This may cause false positives for miRNA promoters. Therefore, more experimental validation is needed to determine the location of the TSSs and promoters of miRNAs in future studies.

In addition to miRNA promoters, numerous *cis*-acting elements of these miRNA promoters have been identified in several species, such as elements regulating plant growth development, hormone response elements, and stress response elements. In the previous study, AtMYC2, ARF, SORLREP3, and LFY transcription-factor-binding site motifs were discovered in *Arabidopsis* to be enriched in miRNA promoters by comparing the promoter elements of 63 miRNA genes and coding genes, as well as randomly selected genomic sequences, using a PWM (Position Weight Matrix) analysis, showing that these transcription factors may be involved in the transcription of *Arabidopsis* miRNAs [40].

PLACE (http://www.dna.affrc.go.jp/htdocs/PLACE, accessed on 12 September 2023) [41] and the PlantCare database (http://bioinformatics.psb.ugent.be/webtools/plantcare/html, accessed on 12 September 2023) [42] are widely used for miRNA promoter *cis*-acting element analyses. In soybean, numerous P-responsive *cis*-elements from the promoters of miRNAs in response to P deficiency were predicted using the PlantCare database [35]. It was found that the frequency of occurrence of the PHR1-binding element, PHO-like-binding element, W-box, and TC element in the promoters of miRNA genes in response to P deficiency (miR156, miR159, miR166, miR167, and miR168, etc.) was higher than that of miRNA genes not responding to P deficiency [35]. In cassava, *cis*-regulatory elements relevant to defense and stress responsiveness, fungal elicitor responsiveness, and hormonal responses were discovered to be abundant in the promoter region of miR160 and miR393 that responds to anthracnose disease, including anaerobic inducible elements (AREs), heat stress response elements (HSE), salicylic acid response elements (TCA-element), TC-rich repeats, fungal inducible response elements (Box-W1), drought-inducible response elements (MBS), and methyl jasmonate response elements (TGACG-motif) [43]. Multiple TC-rich repeats and TCA-elements were also discovered on the promoters of 15 *Arabidopsis* miRNAs responding to *Bacillus velezensis* FZB42 [44].

An analysis of the miRNAs involved in plant salt stress (miR169, miR319, miR393, miR396, and miR398, etc.) in rice revealed that they contained common regulatory elements on their promoters, including GC-boxes, GATA-boxes, MYB response elements, MYC response elements, ABA response elements (ABRE), W-boxes, and zinc finger protein DNA-binding elements (DOF) [45]. Moreover, *cis*-elements for the miRNA genes involved in environmental changes have also been discovered in plants. Environmental SO_2 is a major air pollutant that has a severe impact on plant growth and development. It was found that the regulatory mechanisms of plant miRNAs in response to SO_2 stress have similarities with pathogen-mediated stress responses [46]. An analysis of the promoters of 32 differentially expressed miRNAs in response to SO_2 stress revealed that the fungal-inducer response element Box-W1 and hypoxia response elements (HREs) were more frequently present in the promoters of the SO_2-stress-responsive miRNAs than in the promoters of other miRNAs [46]. miR397, miR398, and miR408 are copper-deficient responsive miRNAs. To investigate the effect of copper concentration on the expression of miRNAs, in vitro cultured grape seedlings were treated with different copper concentrations for 30 days [47]. The miR397a, miR398a, miR398b/c, and miR408 expressions in the grape leaves and roots decreased with an increasing copper concentration [47]. Subsequently, abundant (6–9) CuRE (GTAC core motif) elements were identified on the promoters of four miRNAs, revealing the molecular mechanism of CuRE elements in the plant response to copper deficiency [47].

In addition, *cis*-elements for miRNA genes have been identified in woody plants as well. Multiple hormone response-related elements were identified in the promoters of 13 miRNAs in rubber tree and miR475b in *Populus suaveolens* that responded to low-temperature stress, including the auxin response element (AuxRR-core), gibberellin response element (GARE), salicylic acid response element (TCA-element), ethylene response element (ERE), and jasmonic acid response element (CGTCA-motif, TGACG-motif) [48,49]. In total, 101 classes of *cis*-acting elements were identified in poplar, including abscisic acid response elements (ABREs), heat stress response elements (HSEs), anaerobic-induced elements (AREs), MYB binding sites, low-temperature-induced response elements (LTRs), chloroplast differentiation elements (HD-Zip 1), leaf shape development elements (HD-Zip 2), and endosperm expression elements (GCN4 and Skn-1 motifs) [33].

From these studies, we can see that different stress-responsive miRNAs in different plants have some *cis*-regulatory elements in common and also share some features. The TC-rich element is present in the promoters of the disease-responsive miRNAs of cassava and *Arabidopsis*. W-box is present in the promoters of multiple stress response miRNAs, including P deficiency in soybean, anthracnose disease in cassava, SO_2 stress in *Arabidopsis*, and salt stress in rice. The TCA-element is present in the promoter of the miRNA response to cold in rubber trees and the miRNA response to *Bacillus velezensis* FZB42 in *Arabidopsis*. In the same situation as that for the identification of TSSs and promoters of miRNAs, taking advantage of the biotech software, a large number of *cis*-regulatory elements can be obtained through computational methods. In future studies for certain miRNAs, experimental validation, such as the deletion mutation method, can be applied to determine the core elements of miRNA promoters.

3. Mechanisms of miRNA Transcriptional Regulation

miRNA transcription is regulated by general and specific transcription factors. Transcription factors can bind to *cis*-acting elements on the promoters of miRNAs to activate the transcription of these miRNAs, which is essential for the spatiotemporal-specific expressions of miRNAs or their adaptation to environmental changes. In addition to transcription factors, several other proteins have been found to regulate miRNA transcription by promoting or repressing miRNA transcription directly or indirectly through binding to RNA polymerase II or the miRNA promoter. At present, multiple transcription factors and other proteins have been identified in several species involved in the regulation of multiple biological processes in plant growth and development (Table 2).

Table 2. Summary of upstream transcription factors of miRNAs.

Organism	miRNA	Upstream Transcription Factors of miRNAs	Positive or Negative Regulation of miRNA	Functions of the Modules	Reference
Arabidopsis	miR156	FUS3	Positive	Seed development	[50]
	miR156	ABI3	Positive and negative	Positive regulation in early seed development but negative regulation in late seed development	[51]
	miR775	HY5	Positive and negative	Cell wall remodeling, positive regulation in root growth but negative regulation in aerial organs development	[52]
	miR165/166	HD-ZIP II and III family genes	Negative	Leaf development	[53]
	miR172	AP2, LUG, SEU	Negative	Flower development	[54]
	miR172	PWR	Positive	Flower development	[55]
	miR161 and miR173	AGO1	Positive	Salinity response	[56]
	miR156	PIFs	Negative	Shade response	[57]
	miR163	HY5	Positive	Light response	[58]
	miR399	MYB	Positive	Phosphate starvation response	[59]
	miR160 and miR167	ARF and GARF	Positive	Auxin response	[37]
	miR397b/miR857	EIN2 and EIN3	Positive	Lignin synthesis in response to ethylene signaling	[60]
Arabidopsis and tomato	miR169	HSF	Positive	Heat stress	[61]
Rice	miR156 and miR167h	OsCAMTA4	Positive	Drought response	[62]
Rice	miR396	IDD2	Positive	Stem elongation	[63]
Apple	miR399	SERRATE	Positive	Drought response	[63]
Apple	miR166, miR172 and miR319	SERRATE	Negative	Drought response	[51]
Apple	miR828	MdMYB1	Positive	Anthocyanin synthesis	[64]
Pummelo	miR167a	DREB	Negative	Flower development	[65]
Mulberry	miR172	SPLs	Positive	Flower development	[62]

3.1. The Involvement of Transcriptional Regulation of miRNA in Plant Growth Processes

In *Arabidopsis*, a B3 transcription factor, FUS3, binds to the promoters of *MIR156A* and *MIR156C* and positively regulates the expression levels of pri-miR156a and pri-miR156c [50]. ABI3, an anabolic acid-insensitive protein in the B3 transcription factor family, promotes *MIR156* expression in early seed development, but represses it in late seed development, which is involved in the regulation of the transition from embryo to seedling [51]. The photomorphogenic transcription factor HY5 negatively regulates the expression of *MIR775a* in the aerial organs of *Arabidopsis* and positively regulates its expression in the roots, participating in the process of cell wall remodeling [52]. In rice, the OsIDD2 protein, containing four zinc finger motifs, binds to the *OsmiR396* promoter to promote the transcription of *miR396* and reduce the expression level of its target gene, GRF [63]. Plants overexpressing *OsIDD2* gene show a dwarf phenotype with a higher expression of *OsmiR396* and a lower expression of *GRF1* [63].

3.2. The Involvement of Transcriptional Regulation of miRNA in Plant Leaf Development

The *REVOLUTA (REV)*, *PHABULOSA (PHB)*, and *PHAVOLUTA (PHV)* genes are three *HD-ZIP III* family genes regulating leaf adaxial–abaxial patterning. They are targeted and repressed by abaxially expressed miR165/166 to regulate leaf polarity [66]. However, REV, PHB, and PHV proteins can interact with the HD-ZIP II transcription factors HOMEOBOX ARABIDOPSIS THALIANA 3 (HAT3) and ARABIDOPSIS THALIANA HOMEOBOX 4 (ATHB4) proteins, and the interacting protein complex can bind to conserved *cis*-elements on the *MIR165/166* promoter to repress *MIR165/166* transcription adaxially, which, in turn, represses the expressions of *HD-ZIP III* genes to maintain leaf polarity [53].

3.3. The Involvement of Transcriptional Regulation of miRNA in Plant Flower Development

As a target gene of miR172, the class A gene *APETALA2 (AP2)* is downregulated in inner floral whorl organs such as stamens and carpels [54], while in the outer floral whorl organs of *Arabidopsis*, it has been confirmed that AP2 can recruit LUG, a co-repressor protein of SEU, to the *MIR172* promoter through ChIP, BiFC, yeast two-hybrid, and yeast three-hybrid crosses experiments. Moreover, the miR172 expression is significantly upregulated in *lug*, *seu*, and *ap2* mutants, showing that AP2 can interact with the LEUNIG (LEU) and SEUSS (SEU) proteins to repress miR172 transcription [54]. *SPLs* have been found to be target genes of miR156. In mulberry, six SPL transcription factors recognized the promoter of *MIR172* and activated miR172 expression, revealing that *SPL* genes regulating the transcription of miR172 are involved in the flowering development in perennial woody plants [62]. From a study on citrus flower development, miR167a was found to be specifically expressed in the stamen filaments and anthers of pummelo [65]. The DREB transcription factor can bind to and interact with the *MIR167a* promoter to repress its expression by yeast-one hybrid and dual luciferase assays, revealing the regulatory mechanism of *MIR167* and its upstream element in citrus stamen development [65]. In addition to specific miRNA genes, some transcription factors or proteins can generally bind the promoters of multiple miRNAs. Yeast two-hybrid, pull-down fusion protein sedimentation, and immunoblotting experiments have confirmed that the NOT2 (Negative on TATA less2) protein can bind RNA polymerase II to stimulate miRNA transcription and elongation to regulate miRNA biosynthesis [67]. The expressions of multiple miRNA precursors and mature miRNAs (miR158a, miR159a, miR164b, miR167a, and miR168a) were decreased in *not2* mutants of *Arabidopsis*, leading to severe male organogenesis defects, similar to miRNA mutants [67]. Similar to NOT2, the SANT structural domain protein, the PWR protein, can regulate *MIR172* transcription and floral organ development by promoting the accumulation of RNA polymerase in the promoter regions of *MIR172a* and *MIR172b* [55].

3.4. The Involvement of Transcriptional Regulation of miRNA in the Synthesis of Secondary Metabolites

miR828 plays an important role in the biosynthesis of the anthocyanins in the peel of apple fruit. The expression of miR828 in this peel is maintained at a comparatively low level during the apple fruit coloration stage and increases rapidly during the late trans-color stage [64]. An overexpression of miR828 in apple and *Arabidopsis* decreases anthocyanin synthesis. Yeast one-hybrid and dual luciferase assays have shown that MdMYB1 binds the miR828 promoter and positively regulates miR828 expression, revealing the involvement of MYB-regulated MIR828 transcription in the biosynthesis mechanism of plant fruit anthocyanins [64].

3.5. The Involvement of Transcriptional Regulation of miRNA in Plant Disease Resistance

The *TPR1 (transcriptional corepressor1)* gene is a transcriptional repressor of an *NBS-LRR* gene encoding the disease-resistance protein SNC1 (Suppressor of npr1-1). An overexpression of the *TPR1* gene causes reductions in the levels of several pri-miRNAs and miRNAs (miR164, miR173, miR319, miR390, and miR159) [68]. As a negative regulator of SNC1, the F-box protein CPR1 can mediate the degradation of the SNC1 protein. The *cpr1aba1* mutant results in a large accumulation of the SNC1 protein in the nucleus, causing transcriptional reductions in several miRNAs (pri-miR159a, pri-miR159b, pri-miR164b, pri-miR166a, and pri-miR167a). Since miRNAs (miR173 and miR390, etc.) can target and trigger some disease-resistance genes to produce phasiRNAs, the *cpr1aba1* mutant causes a reduction in the abundance of phasiRNAs produced on four disease-resistance genes, resulting in an upregulation of the expressions of 70 resistance genes. Therefore, miRNA genes are transcriptionally regulated by the *SNC1* gene to participate in plant resistance to pathogens [68].

In addition to the disease resistance originating from the endogenous miRNA targeting and regulation on resistant genes, artificial miRNAs (amiRNAs) and miRNA-induced gene silencing (MIGS) have recently become miRNA-based strategies for obtaining pest and disease resistance [2,69]. Artificial microRNAs (amiRNA) are generally designed from an endogenous miRNA precursor (pre-miRNA), which is used as a structural support in which the miRNA:miRNA* is replaced with a specific miRNA complementary to the desired target sequence [70]. The MIGS approach exploits a special 22-nuclotide miRNA of *Arabidopsis thaliana*, miR173, which can trigger the production of *trans*-acting small RNAs [71]. Different from the miRNA transcription on their own promoter, pre-amiRNAs and the MIGS vector are generally constructed using a plasmid containing an effective constitutive-like *35S* promoter to mediate the targeted viral RNA cleavage to confer resistance to various diseases, such as the Cassava brown streak virus (CBSV), Ugandan cassava brown streak virus (UCBSV), cotton leaf hopper (*Amrasca biguttula*), cotton whitefly (*Bemisia tabaci*), and so on [2,69].

3.6. The Involvement of the Transcriptional Regulation of miRNA in Plant Abiotic Stress

Environmental stresses (such as saline, nutrient deficiency, and heavy metal) greatly constrain normal plant growth and development [72–76]. miRNAs are involved in various abiotic stresses, including salinity, drought, heat, cold, nutrient deficiency, oxidative stress, UV radiation, heavy metal toxicity, and so on [77]. In recent years, the regulation mechanism of miRNAs in the abiotic response has been discovered in some plants. In *Arabidopsis* and tomato, the HSF transcription factor is involved in heat stress tolerance through binding to the *MIR169* promoter to positively regulate the transcription of *MIR169* and negatively regulate the expression of its target gene *NF-YA9/10* [61]. The SERRATE protein, a conserved RNA processing factor in eukaryotes, encodes a C2H2 zinc finger protein. The SERRATE protein can regulate the drought tolerance in apple by positively regulating the transcription of *MIR399* and negatively regulating the transcriptions of *MIR166*, *MIR172*, and *MIR319* [51]. In rice, a Calmodulin-binding Transcription Activator (OsCAMTA4) binds to the promoters of *MIR156* and *MIR167h* to activate the expressions of two miRNAs, participating in the plant response to abiotic stress [62]. In addition to being part of the RNA-silencing complex to cleave mRNA in the cytoplasm, AGO1 also plays roles in miRNA biogenesis at the transcriptional level in the nucleus. From an immunoprecipitation analysis, it was found that AGO1 can bind to the chromatin of miR161 and miR173. In the *ago1* mutant, the expression levels of miR161 and miR173 markedly decreased under sanity stress [56].

Many miRNAs are responsive to environmental signals. The miR156 regulating network is involved in plant adaptation to shade. In shade conditions, the bHLH class protein PHOTOCHROME-INTERACTING FACTORS (PIFs) can bind five *MIR156* promoters, repress their expressions, and concomitantly enhance the expressions of *SPL* family genes to mediate the plant's shade response syndrome (SAS) [57]. The leucine zipper (bZIP)-like transcription factor HY5 (ELONGATED HYPOCOTYL5) can bind to two G/C box elements on the promoter *MIR163* and positively regulate its expression in response to light signals, in order to promote primary root elongation in seedlings without affecting lateral root growth [58].

The MYB transcription factor in *Arabidopsis thaliana* can activate miR399 expression and reduce the expression of its target gene *UBC (ubiquitin-conjugating enzyme)* by binding to the MYB binding site on the *MIR399* promoter in response to phosphate starvation [59]. The restricted expression of UBC relieves the inhibition of the phosphorus transporters by UBC to promote phosphorus uptake and transport [59].

3.7. The Involvement of Transcriptional Regulation of miRNA in Phytohormone Signaling Pathways

Plant hormones play important roles in plant growth and development [78–80]. Several members of the soybean *MIR160* and *MIR167* families contain multiple auxin response factor (ARF)-binding elements and gibberellin response factor (GARF)-binding elements in their promoter regions, suggesting that ARF and GARF transcription factors may bind to the promoters of *MIR160* and *MIR167* to regulate their expressions. Since ARF transcription factor family members such as *ARF17*, *ARF18*, *ARF6*, and *ARF8* are target genes of both miR160 and miR167, a potential feedback regulatory mechanism between miR160 and miR167 and the ARF- and GRF-binding elements was revealed [37]. EIN2 and EIN3 are important transcription factors in the ethylene signaling pathway. An overexpression of *EIN2* in *Arabidopsis* elevates the expression level of miR397b/miR857, as well as reduces the expressions of the target genes *LAC4* and *LAC17*, resulting in a significant reduction in the lignin accumulation in vascular bundles [60]. Yeast one-hybrid experiments have confirmed that EIN2 and EIN3 can bind to ethylene response factors (ERFs) on the *MIR397b/MIR857* promoter to promote transcription, revealing the molecular mechanisms of the miRNAs involved in the regulation of plant lignin synthesis in response to ethylene signaling [60].

3.8. miRNA Transcriptional Regulation Mediated by General Transcription Factors

The transcription of miRNA genes requires the participation of a mediator complex, which can not only recruit RNA polymerase II during transcription, but also interact with specific transcription factors (Figure 2). This mediator complex can bind directly to the miRNA promoters in the presence of some activating proteins to initiate miRNA genes' transcription (Figure 2). In the mediate complex *med20a* mutant, the expressions of six detected miRNA precursors (pri-miR159, pri-miR163, pri-miR165a, pri-miR166a, pri-miR167a, and pri-miR171a) were downregulated by 20%–70% compared to the control and caused abnormal phenotypes such as smaller plants, shorter petioles, the downward bending of leaves, late flowering, and reduced fertility [81,82]. CDC5 (cell division cycle 5), a kind of conserved DNA-binding protein widely found in eukaryotes, is involved in various processes of plant development through binding to the Pol II promoters of multiple miRNAs and then regulating miRNA transcription [83]. In *Arabidopsis cdc5* mutants, several miRNAs (miR166/165, miR167, miR159/319, miR390, miR171, miR172, miR173, miR156, and miR163) were significantly downregulated, resulting in various developmental defects, including plant size, leaf shape, delayed flowering, and sterility [83]. CDF is a kind of

DNA-binding with one finger family protein [84]. *Arabidopsis* CDF2 can bind to miRNA promoters to promote or repress miRNA transcription [85]. In the *cdf2* mutant, two miRNA precursors (pri-miR172a and pri-miR394a) were significantly upregulated, and 17 other miRNA precursors were significantly downregulated [85]. PP4R3A, a regulatory subunit of the phosphatase protein PP4, can bind to Pol II and recruit the polymerase to the promoter regions of miRNAs, thereby promoting miRNA transcription and enhancing the expressions of multiple miRNAs [86]. In addition, SMA1, a homolog of the Prp28 spliceosomal protein, is also required for miRNA transcription [87]. A mutation of the *SMA1* gene resulted in a significant downregulation of 11 miRNA precursors and mature miRNAs and a significant decrease in the Pol II bound to the promoter regions of five miRNAs [87]. In addition, an ATPase chromatin remodeling factor CHR2 and transporter HST can also bind to Pol II to promote the accumulation of polymerase at the miRNA promoter regions [58,87,88].

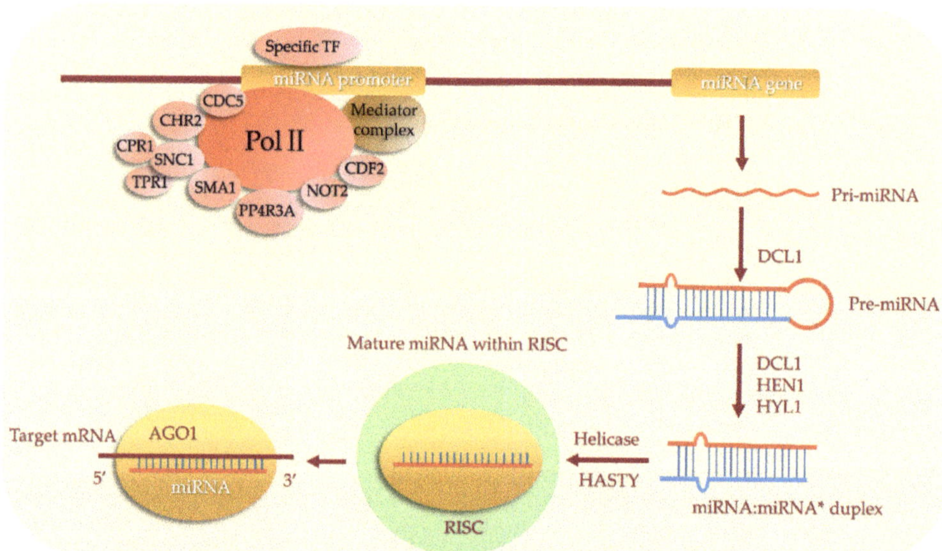

Figure 2. The transcriptional regulation of miRNA by the general transcription factors. *DCL1 (DICER-LIKE 1), HYL1 (methyltransferase HUA ENHANCER 1), HEN1 (HUA ENHANCER 1), AGO1 (ARGONAUTE1), RISC (RNA-induced silencing complex), TPR1 (transcriptional corepressor1), SNC1 (suppressor of npr1-1, constitutive 1), CPR1 (constitutive expresser of PR genes), CDF (Dof transcription factor), SMA1 (Prp28 homolog SMALL 1), CHR (chromatin remodeling factor), CDC5 (CELL DIVISION CYCLE 5), PP4R3A (phosphatase protein PP4), NOT2 (NEGATIVE ON TATA LESS 2), and TF (transcription factor).*

Currently, upstream regulatory elements have been focused on Arabidopsis and few have been identified in other plants. From the present studies, for the same conserved miRNAs, the upstream transcription factors of miRNAs are probably totally different in different developmental stages, such as miR156 in seed development and in leaves (Table 2). And the upstream transcription factors of a same miRNA are probably different in different plant species, such as miR399 in *Arabidopsis* and in apple, demonstrating that miRNAs are tissue-specifically and species-specifically regulated by transcription factors through combining with different cis-elements. Although the mechanisms of miRNA transcriptional regulation are complex, many effective identification methods, such as yeast hybridization, EMSA, ChIP, BiFC, and transgenic, etc., have been shown in previous research and provide a reference for applications in the other miRNAs of different plants.

4. Summary and Prospects

Several studies have been conducted on the transcriptional regulation of miRNA, including multiple miRNAs in model plants and some woody plants, which are involved in multiple biological processes such as plant growth and development, stress response, and signal transduction, etc. However, compared to the number of identified miRNAs, there is still a lack of analysis of the upstream regulators and transcriptional regulation mechanism of miRNA genes. Until now, the transcription factors for only a small number of conserved miRNAs have been demonstrated. How some other conserved and species-specific miRNAs are regulated at the transcriptional level is still unknown. With more and deeper research on the miRNA function in different plants, future research on miRNA promoters and transcriptional regulatory mechanisms will be more extensive. Researchers can identify miRNA promoters and functional elements using 5′ RACE, the GUS staining activity assay, and other techniques, and identify more miRNA upstream regulatory factors using yeast hybridization, EMSA, ChIP, BiFC, and transgenic, etc., to further enrich the study of plant miRNA regulatory mechanisms from upstream to downstream and provide new valuable functional elements and genetic resources for a plant's genetically engineered traits improvement.

Author Contributions: C.T., C.Z., F.G., L.S. and Y.F.; writing—original draft preparation, F.G. and Y.F.; writing—review and editing, Y.F.; supervision. All authors have read and agreed to the published version of the manuscript.

Funding: This research was funded by the National Natural Science Foundation of China (31701891).

Conflicts of Interest: The authors declare no conflict of interest.

References

1. Axtell, M.J. Classification and comparison of small RNAs from plants. *Annu. Rev. Plant Biol.* **2013**, *64*, 137–159. [CrossRef] [PubMed]
2. Hada, A.; Patil, B.L.; Bajpai, A.; Kesiraju, K.; Dinesh-Kumar, S.; Paraselli, B.; Sreevathsa, R.; Rao, U. Micro RNA-induced gene silencing strategy for the delivery of siRNAs targeting *Meloidogyne incognita* in a model plant *Nicotiana benthamiana*. *Pest Manag. Sci.* **2021**, *77*, 3396–3405. [CrossRef] [PubMed]
3. Bartel, D.P. MicroRNAs: Genomics, biogenesis, mechanism, and function. *Cell* **2004**, *116*, 281–297. [CrossRef] [PubMed]
4. Gao, Z.; Nie, J.; Wang, H. MicroRNA biogenesis in plant. *Plant Growth Regul.* **2021**, *93*, 1–12. [CrossRef]
5. Wang, J.; Mei, J.; Ren, G. Plant microRNAs: Biogenesis, homeostasis, and degradation. *Front. Plant Sci.* **2019**, *10*, 360. [CrossRef]
6. Voinnet, O. Origin, biogenesis, and activity of plant microRNAs. *Cell* **2009**, *136*, 669–687. [CrossRef]
7. Liu, H.; Yu, H.; Tang, G.; Huang, T. Small but powerful: Function of microRNAs in plant development. *Plant Cell Rep.* **2018**, *37*, 515–528. [CrossRef]
8. Spanudakis, E.; Jackson, S. The role of microRNAs in the control of flowering time. *J. Exp. Bot.* **2014**, *65*, 365–380. [CrossRef] [PubMed]
9. Chen, Q.; Wang, J.; Danzeng, P.; Danzeng, C.; Song, S.; Wang, L.; Zhao, L.; Xu, W.; Zhang, C.; Ma, C. VvMYB114 mediated by miR828 negatively regulates trichome development of *Arabidopsis*. *Plant Sci.* **2021**, *309*, 110936. [CrossRef]
10. Barrera-Rojas, C.H.; Rocha, G.H.B.; Polverari, L.; Pinheiro Brito, D.A.; Batista, D.S.; Notini, M.M.; da Cruz, A.C.F.; Morea, E.G.O.; Sabatini, S.; Otoni, W.C. miR156-targeted SPL10 controls Arabidopsis root meristem activity and root-derived *de novo* shoot regeneration via cytokinin responses. *J. Exp. Bot.* **2020**, *71*, 934–950. [CrossRef]
11. Samad, A.F.; Sajad, M.; Nazaruddin, N.; Fauzi, I.A.; Murad, A.M.; Zainal, Z.; Ismail, I. MicroRNA and transcription factor: Key players in plant regulatory network. *Front. Plant Sci.* **2017**, *8*, 565. [CrossRef] [PubMed]
12. Dong, Q.; Hu, B.; Zhang, C. microRNAs and their roles in plant development. *Front. Plant Sci.* **2022**, *13*, 824240. [CrossRef] [PubMed]
13. Waheed, S.; Anwar, M.; Saleem, M.A.; Wu, J.; Tayyab, M.; Hu, Z. The critical role of small RNAs in regulating plant innate immunity. *Biomolecules* **2021**, *11*, 184. [CrossRef] [PubMed]
14. Wang, R.; Fang, Y.-N.; Wu, X.-M.; Qing, M.; Li, C.-C.; Xie, K.-D.; Deng, X.-X.; Guo, W.-W. The miR399-CsUBC24 module regulates reproductive development and male fertility in citrus. *Plant Physiol.* **2020**, *183*, 1681–1695. [CrossRef] [PubMed]
15. Das, R.; Mukherjee, A.; Basak, S.; Kundu, P. Plant miRNA responses under temperature stress. *Plant Gene* **2021**, *28*, 100317. [CrossRef]
16. Gao, Z.; Ma, C.; Zheng, C.; Yao, Y.; Du, Y. Advances in the regulation of plant salt-stress tolerance by miRNA. *Mol. Biol. Rep.* **2022**, *49*, 5041–5055. [CrossRef]

17. Li, S.; Gao, F.; Xie, K.; Zeng, X.; Cao, Y.; Zeng, J.; He, Z.; Ren, Y.; Li, W.; Deng, Q. The OsmiR396c-OsGRF4-OsGIF1 regulatory module determines grain size and yield in rice. *Plant Biotechnol. J.* **2016**, *14*, 2134–2146. [CrossRef]
18. Gao, F.; Wang, K.; Liu, Y.; Chen, Y.; Chen, P.; Shi, Z.; Luo, J.; Jiang, D.; Fan, F.; Zhu, Y. Blocking miR396 increases rice yield by shaping inflorescence architecture. *Nat. Plants* **2015**, *2*, 15196. [CrossRef]
19. Sun, X.; Lin, L.; Sui, N. Regulation mechanism of microRNA in plant response to abiotic stress and breeding. *Mol. Biol. Rep.* **2019**, *46*, 1447–1457. [CrossRef]
20. Lee, Y.; Lee, D.; Cho, L.; An, G. Rice *miR172* induces flowering by suppressing *OsIDS1* and *SNB*, two AP2 genes that negatively regulate expression of *Ehd1* and florigens. *Rice* **2014**, *7*, 31. [CrossRef]
21. Jodder, J. miRNA-mediated regulation of auxin signaling pathway during plant development and stress responses. *J. Biosci.* **2020**, *45*, 91. [CrossRef] [PubMed]
22. Liu, N.; Wu, S.; Van Houten, J.; Wang, Y.; Ding, B.; Fei, Z.; Clarke, T.H.; Reed, J.W.; Van Der Knaap, E. Down-regulation of AUXIN RESPONSE FACTORS 6 and 8 by microRNA 167 leads to floral development defects and female sterility in tomato. *J. Exp. Bot.* **2014**, *65*, 2507–2520. [CrossRef] [PubMed]
23. Li, Y.; Chen, T.; Khan, W.U.; An, X. Regulatory roles of miRNAs associated with the aging pathway in tree vegetative phase changes. *For. Res.* **2023**, *3*, 9. [CrossRef]
24. Bajczyk, M.; Jarmolowski, A.; Jozwiak, M.; Pacak, A.; Pietrykowska, H.; Sierocka, I.; Swida-Barteczka, A.; Szewc, L.; Szweykowska-Kulinska, Z. Recent insights into plant miRNA biogenesis: Multiple layers of miRNA level regulation. *Plants* **2023**, *12*, 342. [CrossRef] [PubMed]
25. Yu, Y.; Jia, T.R.; Chen, X.M. The "how" and "where" of plant microRNAs. *New Phytol.* **2017**, *216*, 1002–1017. [CrossRef] [PubMed]
26. Zhou, X.; Ruan, J.; Wang, G.; Zhang, W. Characterization and identification of microRNA core promoters in four model species. *PLoS Comput. Biol.* **2007**, *3*, e37. [CrossRef] [PubMed]
27. Borsani, O.; Zhu, J.; Verslues, P.E.; Sunkar, R.; Zhu, J.-K. Endogenous siRNAs derived from a pair of natural *cis*-antisense transcripts regulate salt tolerance in *Arabidopsis*. *Cell* **2005**, *123*, 1279–1291. [CrossRef]
28. Jouannet, V.; Maizel, A. *Trans*-acting small interfering RNAs: Biogenesis, mode of action, and role in plant development. *MicroRNAs Plant Dev. Stress Responses* **2012**, *15*, 83–108.
29. Blevins, T.; Pontes, O.; Pikaard, C.S.; Meins, F., Jr. Heterochromatic siRNAs and DDM1 independently silence aberrant 5S rDNA transcripts in *Arabidopsis*. *PLoS ONE* **2009**, *4*, e5932. [CrossRef]
30. Xie, Z.; Allen, E.; Fahlgren, N.; Calamar, A.; Givan, S.A.; Carrington, J.C. Expression of Arabidopsis *MIRNA* genes. *Plant Physiol.* **2005**, *138*, 2145–2154. [CrossRef]
31. Dai, Z.; Gao, J.; An, K.; Lee, J.M.; Edwards, G.E.; An, G. Promoter elements controlling developmental and environmental regulation of a tobacco ribosomal protein gene L34. *Plant Mol. Biol.* **1996**, *32*, 1055–1065. [CrossRef] [PubMed]
32. Cui, X.; Xu, S.M.; Mu, D.S.; Yang, Z.M. Genomic analysis of rice microRNA promoters and clusters. *Gene* **2009**, *431*, 61–66. [CrossRef] [PubMed]
33. Chen, M.; Wei, M.; Dong, Z.; Bao, H.; Wang, Y. Genomic identification of microRNA promoters and their *cis*-acting elements in *Populus*. *Genes Genom.* **2016**, *38*, 377–387. [CrossRef]
34. Liu, Y.-X.; Han, Y.-P.; Chang, W.; Quan, Z.; Guo, M.-Z.; Li, W.-B. Genomic analysis of microRNA promoters and their *cis*-acting elements in soybean. *Agric. Sci. China* **2010**, *9*, 1561–1570. [CrossRef]
35. Zeng, H.Q.; Zhu, Y.Y.; Huang, S.Q.; Yang, Z.M. Analysis of phosphorus-deficient responsive miRNAs and *cis*-elements from soybean (*Glycine max* L.). *J. Plant Physiol.* **2010**, *167*, 1289–1297. [CrossRef]
36. Zhao, Y.; Wang, F.; Juan, L. MicroRNA promoter identification in arabidopsis using multiple histone markers. *BioMed Res. Int.* **2015**, *2015*, 861402. [CrossRef] [PubMed]
37. Han, Y.-Q.; Hu, Z.; Zheng, D.-F.; Gao, Y.-M. Analysis of promoters of microRNAs from a *Glycine max* degradome library. *J. Zhejiang Univ. Sci. B* **2014**, *15*, 125–132. [CrossRef]
38. Sriwichai, N.; Saithong, T.; Thammarongtham, C.; Meechai, A.; Kalapanulak, S. A hybrid computational approach for predicting the intergenic microRNA promoters in plants: A case study in cassava. In Proceedings of the 27th Annual Meeting of the Thai Society for Biotechnology and International Conference, Mandarin Hotel Bangkok by Centre Point, Bangkok, Thailand, 17–20 November 2015.
39. Zhou, X.; Ruan, J.; Wang, G.; Zhang, W. Characterization of the promoters of microRNA genes: A genome-scale analysis on *C. elegans*, *A. thaliana* and *H. sapiens*. In Proceedings of the First Annual RECOMB Satellite Workshop on Systems Biology and the Second Annual RECOMB Satellite Workshop on Regulatory Genomics, San Diego, CA, USA, 2–4 December 2005.
40. Megraw, M.; Baev, V.; Rusinov, V.; Jensen, S.T.; Kalantidis, K.; Hatzigeorgiou, A.G. MicroRNA promoter element discovery in *Arabidopsis*. *RNA* **2006**, *12*, 1612–1619. [CrossRef]
41. Higo, K.; Ugawa, Y.; Iwamoto, M.; Higo, H. PLACE: A database of plant *cis*-acting regulatory DNA elements. *Nucleic Acids Res.* **1998**, *26*, 358–359. [CrossRef]
42. Lescot, M.; Déhais, P.; Thijs, G.; Marchal, K.; Moreau, Y.; Van de Peer, Y.; Rouzé, P.; Rombauts, S. PlantCARE, a database of plant *cis*-acting regulatory elements and a portal to tools for *in silico* analysis of promoter sequences. *Nucleic Acids Res.* **2002**, *30*, 325–327. [CrossRef]
43. Pinweha, N.; Asvarak, T.; Viboonjun, U.; Narangajavana, J. Involvement of miR160/miR393 and their targets in cassava responses to anthracnose disease. *J. Plant Physiol.* **2015**, *174*, 26–35. [CrossRef] [PubMed]

44. Xie, S.; Jiang, H.; Xu, Z.; Xu, Q.; Cheng, B. Small RNA profiling reveals important roles for miRNAs in *Arabidopsis* response to *Bacillus velezensis* FZB42. *Gene* **2017**, *629*, 9–15. [CrossRef] [PubMed]
45. Haldar, S.; Bandyopadhyay, S. Co-ordinated regulation of miRNA and their target genes by CREs during salt stress in *Oryza sativa* (Rice). *Plant Gene* **2021**, *28*, 100323. [CrossRef]
46. Li, L.; Xue, M.; Yi, H. Uncovering microRNA-mediated response to SO_2 stress in *Arabidopsis thaliana* by deep sequencing. *J. Hazard. Mater.* **2016**, *316*, 178–185. [CrossRef] [PubMed]
47. Leng, X.; Wang, P.; Zhao, P.; Wang, M.; Cui, L.; Shangguan, L.; Wang, C. Conservation of microRNA-mediated regulatory networks in response to copper stress in grapevine. *Plant Growth Regul.* **2017**, *82*, 293–304. [CrossRef]
48. Niu, J.; Wang, J.; Hu, H.; Chen, Y.; An, J.; Cai, J.; Sun, R.; Sheng, Z.; Liu, X.; Lin, S. Cross-talk between freezing response and signaling for regulatory transcriptions of MIR475b and its targets by miR475b promoter in *Populus suaveolens*. *Sci. Rep.* **2016**, *6*, 20648. [CrossRef] [PubMed]
49. Kanjanawattanawong, S.; Tangphatsornruang, S.; Triwitayakorn, K.; Ruang-areerate, P.; Sangsrakru, D.; Poopear, S.; Somyong, S.; Narangajavana, J. Characterization of rubber tree microRNA in phytohormone response using large genomic DNA libraries, promoter sequence and gene expression analysis. *Mol. Genet. Genom.* **2014**, *289*, 921–933. [CrossRef]
50. Wang, F.; Perry, S.E. Identification of direct targets of FUSCA3, a key regulator of Arabidopsis seed development. *Plant Physiol.* **2013**, *161*, 1251–1264. [CrossRef] [PubMed]
51. Tian, R.; Wang, F.; Zheng, Q.; Niza, V.M.; Downie, A.B.; Perry, S.E. Direct and indirect targets of the *Arabidopsis* seed transcription factor ABSCISIC ACID INSENSITIVE3. *Plant J.* **2020**, *103*, 1679–1694. [CrossRef]
52. Zhang, H.; Guo, Z.; Zhuang, Y.; Suo, Y.; Du, J.; Gao, Z.; Pan, J.; Li, L.; Wang, T.; Xiao, L. MicroRNA775 regulates intrinsic leaf size and reduces cell wall pectin levels by targeting a galactosyltransferase gene in *Arabidopsis*. *Plant Cell* **2021**, *33*, 581–602. [CrossRef]
53. Merelo, P.; Ram, H.; Pia Caggiano, M.; Ohno, C.; Ott, F.; Straub, D.; Graeff, M.; Cho, S.K.; Yang, S.W.; Wenkel, S. Regulation of MIR165/166 by class II and class III homeodomain leucine zipper proteins establishes leaf polarity. *Proc. Natl. Acad. Sci. USA* **2016**, *113*, 11973–11978. [CrossRef] [PubMed]
54. Grigorova, B.; Mara, C.; Hollender, C.; Sijacic, P.; Chen, X.; Liu, Z. LEUNIG and SEUSS co-repressors regulate miR172 expression in *Arabidopsis* flowers. *Development* **2011**, *138*, 2451–2456. [CrossRef] [PubMed]
55. Yumul, R.E.; Kim, Y.J.; Liu, X.; Wang, R.; Ding, J.; Xiao, L.; Chen, X. POWERDRESS and diversified expression of the MIR172 gene family bolster the floral stem cell network. *PLoS Genet.* **2013**, *9*, e1003218. [CrossRef] [PubMed]
56. Dolata, J.; Bajczyk, M.; Bielewicz, D.; Niedojadlo, K.; Niedojadlo, J.; Pietrykowska, H.; Walczak, W.; Szweykowska-Kulinska, Z.; Jarmolowski, A. Salt stress reveals a new role for ARGONAUTE1 in miRNA biogenesis at the transcriptional and posttranscriptional levels. *Plant Physiol.* **2016**, *172*, 297–312. [CrossRef] [PubMed]
57. Xie, Y.; Liu, Y.; Wang, H.; Ma, X.; Wang, B.; Wu, G.; Wang, H. Phytochrome-interacting factors directly suppress MIR156 expression to enhance shade-avoidance syndrome in *Arabidopsis*. *Nat. Commun.* **2017**, *8*, 348. [CrossRef]
58. Cambiagno, D.A.; Giudicatti, A.J.; Arce, A.L.; Gagliardi, D.; Li, L.; Yuan, W.; Lundberg, D.S.; Weigel, D.; Manavella, P.A. HASTY modulates miRNA biogenesis by linking pri-miRNA transcription and processing. *Mol. Plant* **2021**, *14*, 426–439. [CrossRef]
59. Baek, D.; Kim, M.C.; Chun, H.J.; Kang, S.; Park, H.C.; Shin, G.; Park, J.; Shen, M.; Hong, H.; Kim, W.-Y. Regulation of miR399f transcription by AtMYB2 affects phosphate starvation responses in *Arabidopsis*. *Plant Physiol.* **2013**, *161*, 362–373. [CrossRef]
60. Gaddam, S.R.; Bhatia, C.; Gautam, H.; Pathak, P.K.; Sharma, A.; Saxena, G.; Trivedi, P.K. Ethylene regulates miRNA-mediated lignin biosynthesis and leaf serration in *Arabidopsis thaliana*. *Biochem. Biophys. Res. Commun.* **2022**, *605*, 51–55. [CrossRef]
61. Rao, S.; Gupta, A.; Bansal, C.; Sorin, C.; Crespi, M.; Mathur, S. A conserved HSF: miR169: NF-YA loop involved in tomato and *Arabidopsis* heat stress tolerance. *Plant J.* **2022**, *112*, 7–26. [CrossRef]
62. Kansal, S.; Panwar, V.; Mutum, R.D.; Raghuvanshi, S. Investigations on regulation of micrornas in rice reveal $[Ca^{2+}]_{cyt}$ signal transduction regulated MicroRNAs. *Front. Plant Sci.* **2021**, *12*, 720009. [CrossRef]
63. Lu, Y.; Feng, Z.; Meng, Y.; Bian, L.; Xie, H.; Mysore, K.S.; Liang, J. SLENDER RICE1 and *Oryza sativa* INDETERMINATE DOMAIN2 regulating OsmiR396 are involved in stem elongation. *Plant Physiol.* **2020**, *182*, 2213–2227. [CrossRef] [PubMed]
64. Li, X.; Chen, P.; Xie, Y.; Yan, Y.; Wang, L.; Dang, H.; Zhang, J.; Xu, L.; Ma, F.; Guan, Q. Apple SERRATE negatively mediates drought resistance by regulating MdMYB88 and MdMYB124 and microRNA biogenesis. *Hortic. Res.* **2020**, *7*, 98. [CrossRef] [PubMed]
65. Fang, Y.-N.; Zheng, B.-B.; Wang, L.; Yang, W.; Wu, X.-M.; Xu, Q.; Guo, W.-W. High-throughput sequencing and degradome analysis reveal altered expression of miRNAs and their targets in a male-sterile cybrid pummelo (*Citrus grandis*). *BMC Genom.* **2016**, *17*, 591. [CrossRef] [PubMed]
66. Zhong, R.; Ye, Z.-H. Regulation of HD-ZIP III genes by microRNA 165. *Plant Signal. Behav.* **2007**, *2*, 351–353. [CrossRef]
67. Wang, L.; Song, X.; Gu, L.; Li, X.; Cao, S.; Chu, C.; Cui, X.; Chen, X.; Cao, X. NOT2 proteins promote polymerase II–dependent transcription and interact with multiple microRNA biogenesis factors in *Arabidopsis*. *Plant Cell* **2013**, *25*, 715–727. [CrossRef] [PubMed]
68. Cai, Q.; Liang, C.; Wang, S.; Hou, Y.; Gao, L.; Liu, L. The disease resistance protein SNC1 represses the biogenesis of microRNAs and phased siRNAs. *Nat. Commun.* **2018**, *9*, 5080. [CrossRef]
69. Wagaba, H.; Patil, B.L.; Mukasa, S.; Alicai, T.; Fauquet, C.M.; Taylor, N.J. Artificial microRNA-derived resistance to Cassava brown streak disease. *J. Virol. Methods* **2016**, *231*, 38–43. [CrossRef]

70. Ai, T.; Zhang, L.; Gao, Z.; Zhu, C.; Guo, X. Highly efficient virus resistance mediated by artificial microRNAs that target the suppressor of PVX and PVY in plants. *Plant Biol.* **2011**, *13*, 304–316. [CrossRef]
71. de Felippes, F.F.; Wang, J.w.; Weigel, D. MIGS: miRNA-induced gene silencing. *Plant J.* **2012**, *70*, 541–547. [CrossRef]
72. Luo, J.; Liang, Z.; Wu, M.; Mei, L. Genome-wide identification of BOR genes in poplar and their roles in response to various environmental stimuli. *Environ. Exp. Bot.* **2019**, *164*, 101–113. [CrossRef]
73. Luo, J.; Shi, W.; Li, H.; Janz, D.; Luo, Z.-B. The conserved salt-responsive genes in the roots of *Populus* × *canescens* and *Arabidopsis thaliana*. *Environ. Exp. Bot.* **2016**, *129*, 48–56. [CrossRef]
74. Ren, H.; Zhong, Y.; Guo, L.; Hussian, J.; Zhou, C.; Cao, Y.; Wu, W.; Liu, S.; Qi, G. Molecular mechanisms of low-temperature sensitivity in tropical/subtropical plants: A case study of *Casuarina equisetifolia*. *For. Res.* **2023**, *3*, 20. [CrossRef]
75. Tong, R.; Wen, Y.; Wang, J.; Lou, C.; Ma, C.; Zhu, N.; Yuan, W.; Geoff Wang, G.; Wu, T. Root nutrient capture and leaf resorption efficiency modulated by different influential factors jointly alleviated P limitation in *Quercus acutissima* across the North–South Transect of Eastern China. *For. Res.* **2022**, *2*, 7. [CrossRef]
76. Luo, J.; Zhou, J.-J. Growth performance, photosynthesis, and root characteristics are associated with nitrogen use efficiency in six poplar species. *Environ. Exp. Bot.* **2019**, *164*, 40–51. [CrossRef]
77. Begum, Y. Regulatory role of microRNAs (miRNAs) in the recent development of abiotic stress tolerance of plants. *Gene* **2022**, *821*, 146283. [CrossRef] [PubMed]
78. Luo, J.; Zhou, J.-J.; Zhang, J.-Z. *Aux/IAA* gene family in plants: Molecular structure, regulation, and function. *Int. J. Mol. Sci.* **2018**, *19*, 259. [CrossRef] [PubMed]
79. Zhou, J.-J.; Luo, J. The PIN-FORMED auxin efflux carriers in plants. *Int. J. Mol. Sci.* **2018**, *19*, 2759. [CrossRef]
80. Luo, J.; Xia, W.; Cao, P.; Xiao, Z.; Zhang, Y.; Liu, M.; Zhan, C.; Wang, N. Integrated transcriptome analysis reveals plant hormones jasmonic acid and salicylic acid coordinate growth and defense responses upon fungal infection in poplar. *Biomolecules* **2019**, *9*, 12. [CrossRef]
81. Tsai, K.-L.; Tomomori-Sato, C.; Sato, S.; Conaway, R.C.; Conaway, J.W.; Asturias, F.J. Subunit architecture and functional modular rearrangements of the transcriptional mediator complex. *Cell* **2014**, *157*, 1430–1444. [CrossRef]
82. Kim, Y.J.; Zheng, B.; Yu, Y.; Won, S.Y.; Mo, B.; Chen, X. The role of Mediator in small and long noncoding RNA production in *Arabidopsis thaliana*. *EMBO J.* **2011**, *30*, 814–822. [CrossRef]
83. Zhang, S.; Xie, M.; Ren, G.; Yu, B. CDC5, a DNA binding protein, positively regulates posttranscriptional processing and/or transcription of primary microRNA transcripts. *Proc. Natl. Acad. Sci. USA* **2013**, *110*, 17588–17593. [CrossRef] [PubMed]
84. Gupta, S.; Malviya, N.; Kushwaha, H.; Nasim, J.; Bisht, N.C.; Singh, V.; Yadav, D. Insights into structural and functional diversity of Dof (DNA binding with one finger) transcription factor. *Planta* **2015**, *241*, 549–562. [CrossRef]
85. Sun, Z.; Guo, T.; Liu, Y.; Liu, Q.; Fang, Y. The roles of Arabidopsis CDF2 in transcriptional and posttranscriptional regulation of primary microRNAs. *PLoS Genet.* **2015**, *11*, e1005598.
86. Wang, S.; Quan, L.; Li, S.; You, C.; Zhang, Y.; Gao, L.; Zeng, L.; Liu, L.; Qi, Y.; Mo, B. The PROTEIN PHOSPHATASE4 complex promotes transcription and processing of primary microRNAs in *Arabidopsis*. *Plant Cell* **2019**, *31*, 486–501. [CrossRef] [PubMed]
87. Li, S.; Xu, R.; Li, A.; Liu, K.; Gu, L.; Li, M.; Zhang, H.; Zhang, Y.; Zhuang, S.; Wang, Q. SMA1, a homolog of the splicing factor Prp28, has a multifaceted role in miRNA biogenesis in *Arabidopsis*. *Nucleic Acids Res.* **2018**, *46*, 9148–9159. [CrossRef]
88. Li, M.; Yu, B. Recent advances in the regulation of plant miRNA biogenesis. *RNA Biol.* **2021**, *18*, 2087–2096. [CrossRef]

life

Article

ZmCOP1 Regulates Maize Mesocotyl Length and Plant Height through the Phytohormone Pathways

Liping Chen [1,2,†], Qiuhua Li [1,2,†], Ming Wang [1,2,†], Feng Xiao [1,2], Kangshi Li [1,2], Ran Yang [1,2], Meng Sun [1,2], Haiyan Zhang [1,2], Jinjie Guo [3,*], Jingtang Chen [1,2,*] and Fuchao Jiao [1,2,*]

1 College of Agronomy, Qingdao Agricultural University, Qingdao 266109, China
2 The Characteristic Laboratory of Crop Germplasm Innovation and Application, Provincial Department of Education, College of Agronomy, Qingdao Agricultural University, Qingdao 266109, China
3 State Key Laboratory of North China Crop Improvement and Regulation, Hebei Sub-Center for National Maize Improvement Center, College of Agronomy, Hebei Agricultural University, Baoding 071001, China
* Correspondence: guojinjie512@163.com (J.G.); chenjingtang@126.com (J.C.); fuchao.jiao@qau.edu.cn (F.J.)
† These authors contributed equally to this work.

Abstract: The morphogenesis of crops is critical to their yield performance. COP1 (constitutively photomorphogenic1) is one of the core regulators in plant morphogenesis and has been deeply studied in *Arabidopsis thaliana*. However, the function of COP1 in maize is still unclear. Here, we found that the mesocotyl lengths of *zmcop1* loss-of-function mutants were shorter than those of wild-type B73 in darkness, while the mesocotyl lengths of lines with *ZmCOP1* overexpression were longer than those of wild-type B104. The plant height with *zmcop1* was shorter than that of B73 in both short- and long-day photoperiods. Using transcriptome RNA sequencing technology, we identified 33 DEGs (differentially expressed genes) between B73′s etiolated seedlings and those featuring *zmcop1*, both in darkness. The DEGs were mainly enriched in the plant phytohormone pathways. Our results provide direct evidence that *ZmCOP1* functions in the elongation of etiolated seedlings in darkness and affects plant height in light. Our data can be applied in the improvement of maize plant architecture.

Keywords: *ZmCOP1*; mesocotyl elongation; plant height; RNA sequencing; phytohormone

Citation: Chen, L.; Li, Q.; Wang, M.; Xiao, F.; Li, K.; Yang, R.; Sun, M.; Zhang, H.; Guo, J.; Chen, J.; et al. *ZmCOP1* Regulates Maize Mesocotyl Length and Plant Height through the Phytohormone Pathways. *Life* **2023**, *13*, 1522. https://doi.org/10.3390/life13071522

Academic Editors: Jie Luo and Sen Meng

Received: 10 April 2023
Revised: 27 June 2023
Accepted: 30 June 2023
Published: 7 July 2023

1. Introduction

Plants exhibit different morphological characteristics when grown in darkness and in light; together, these phenomenon is called morphogenesis [1,2]. The elongation of the hypocotyl or mesocotyl is critical to seedling emergence [3–5]. In *Arabidopsis thaliana*, COP1 (constitutively photomorphogenic 1) plays a central role in morphogenesis [6–8]. In darkness, the *Arabidopsis thaliana* hypocotyl elongates, while weak *cop1* mutants have shown a short hypocotyl length and open cotyledons [9].

In darkness, COP1 usually binds SPA1 (suppressor of phya 1) to form the COP1–SPA complex [10]. PIF1 (phytochrome interacting factor 1) interacts with the COP1–SPA complex to induce degradation of HY5 (elongated hypocotyl 5), resulting in hypocotyl elongation [11–14]. COP1 also participates in the degradation of WDL3 (wave-dampened 2-like 3) through the 26S proteasome-mediated pathway, leading to hypocotyl elongation [1,15].

In light, CRY (cryptochrome) inhibits the regulation activity of the COP1–SPA complex and stabilizes HY5 [16–20]. The Pfr (far-red light-absorbing form) of the photosensitive pigment induces phosphorylation of PIFs. Phosphorylated PIFs are recognized by COP1–SPA, then ubiquitinated and degraded by the 26S proteasome [14,21]. Thus, the stability of HY5 and the degradation of PIFs inhibit hypocotyl elongation. COP1 is also involved in the regulation of the circadian clock (Bhatnagar et al., 2020). A weak *cop1* mutant showed a short circadian clock gene expression cycle and an early flowering phenotype under short daylight [9,15,22–24].

Hormones such as BR (brassinolide), JA (jasmonic acid) and ETH (ethylene) play important roles in the regulation of morphogenesis. In *Arabidopsis*, BR promotes skotomorphogenesis. BRI1 (BR-insensitive1) and BIN2 (BR-insensitive2) promote accumulation of BBX28 and BBX29. BBX28 and BBX29 interact with BEE1 (BR-enhanced expression1), BEE2 (BR-enhanced expression2) and BEE3 (BR-enhanced expression3) [25]. HY5 enhances the activity of GSK3-like kinase BIN2 (brassinosteroid-insensitive 2) to repress skotomorphogenesis [24]. Ethylene inhibits EBF1 (ethylene response factor1) and EBF2 (ethylene response factor2) to stabilize EIN3 (ethylene-insensitive 3) and EIL1 (EIN3-like 1), respectively, inhibiting the opening and expansion of Arabidopsis cotyledons and maintaining skotomorphogenesis [26,27].

COP1 homologs have been identified in various plants, including *Arabidopsis thaliana, Sorghum bicolor, Zea mays* and *Oryza sativa* (Huai et al., 2020). Although *ZmCOP1* has been shown to restore the *atcop1-4* phenotypes in Arabidopsis, the function of *ZmCOP1* in maize has not been well studied. Here, we have phenotypically characterized *zmcop1* mutants and overexpression lines and identified DEGs between *zmcop1* mutants and wild types using RNA sequencing technology. We have shown that *ZmCOP1* may inhibit the elongation of the mesocotyl through the *BR* signal transduction pathway. We have provided evidence that *ZmCOP1* has a conserved function in *Zea mays* and *Arabidopsis thaliana*.

2. Materials and Methods

2.1. Plant Material and Growth Conditions

zmcop1-1 and *zmcop1-2* mutants were collected from the maize EMS mutant library (www.elabcaas.cn, accessed on 14 September 2021). The wild type (B73 inbred line) from the same library was used as a control.

For the wild type (WT), *zmcop1-1* and *zmcop1-2*, a mesocotyl elongation analysis and an RT-qPCR analysis were performed. In a chi-square test, the WT and *zmcop1-1* were crossed to generate F_1 hybrid lines; then, F_1 hybrid seeds were self-fertilized to generate F_2 lines. Around 140 seeds from one F_2 ear were planted in soil in a greenhouse. The F_2 seedlings were grown in the dark for 5–7 days. For plant height analysis, WT and *zmcop1-1* were planted in Jiaozhou (36.26429, 120.03192) in the summer of 2021, in Ledong (36.26429, 120.03192) in the winter of 2021 and in Jiaozhou again in the summer of 2022. The mesocotyl length and the plant height of the 7 day old etiolated seedlings were measured according to the standard methods.

ZmCOP1 overexpression transgenic lines were generated by Beijing Bomei Xing'ao Technology Co., Ltd. In general, the *ZmCOP1* coding sequence was amplified and inserted into the 521 plasmid, furthered by the *ZmUBI* promoter. The transformation was performed following the standard Agrobacteria-mediated transformation protocol for maize, using B104 immature embryos. Positive transformation events were selected based on kanamycin and bar herbicide resistance. Positive transgenic lines were confirmed with PCR.

2.2. DNA Extraction and Genotyping

The CTAB method was used to extract the total DNA from leaves grown for 7 days. Two pairs of primers were designed near two mutation sites using primer5.0. To perform PCR, 2x Taq PCR StarMix with Loading Dye (GenStar) was used. A 1% agarose gel electrophoresis experiment was used for the last step of genotype detection.

2.3. Measurement of SPAD Value

The SPAD values (relative content of chlorophyll) of the ear leaf and the first leaf between the WT and *zmcop1* were measured with SPAD502 (Zhejiang Top Cloud-Agri Technology Co., Ltd., Hangzhou, China). The measuring position of the ear leaf was about 6 cm from the base of the first ear, and the measuring site of the first leaf was about 10 cm from the tip of the leaf. At least 15 groups of data were collected.

2.4. RNA Sequencing

2.4.1. RNA Extraction

The construction of a cDNA library and an RNA sequencing analysis was completed with Wuhan MetWare Biotechnology Co., Ltd. (Wuhan, China).

Seven-day-old etiolated seedlings were collected for RNA isolation. Four biological replicates were used. The RNA purity was checked using a NanoPhotometer® spectrophotometer (IMPLEN, Westlake Village, CA, USA). The RNA concentration was measured using a Qubit®2.0 Fluorometer (Thermo Fisher Scientific, Carlsbad, CA, USA). The RNA integrity was assessed using an RNA Nano 6000 Assay Kit of a Bioanalyzer 2100 system (Agilent Technologies, Santa Clara, CA, USA).

2.4.2. Library Preparation for Transcriptome Sequencing

A total of 1 μg of RNA was used for each sample library preparation. Sequencing libraries were generated using a NEBNext® Ultra™ RNA Library Prep Kit for Illumina® (New England Biolabs, Ipswich, MA, USA), following the manufacturer's recommendations. Briefly, mRNA was purified from the total RNA using poly-T oligo attached magnetic beads. Fragmentation was carried out using divalent cations and under elevated temperatures in NEBNext First Strand Synthesis Reaction Buffer (5X). The first-strand cDNA was synthesized using a random hexamer primer and M-MuLV Reverse Transcriptase (RNase H-). Second-strand cDNA synthesis was performed using DNA Polymerase I and RNase H. cDNA fragments (preferentially 250~300 bp in length) were purified with an AMPure XP system (Beckman Coulter, Indianapolis, IN, USA). The PCR products were purified (AMPure XP system), and the library quality was assessed with the Agilent Bioanalyzer 2100 system.

2.4.3. Clustering and Sequencing

Clustering of the index-coded samples was performed with a cBot Cluster Generation System using a TruSeq PE Cluster Kit v3-cBot-HS (Illumia, San Diego, CA, USA), according to the manufacturer's instructions. After cluster generation, the library preparations were sequenced on an Illumina platform and 150 bp paired-end reads were generated.

2.4.4. Analysis of the RNA-Seq Data

The raw sequencing data were filtered using fastp v0.19.3 with adapters and then aligned to Zm-B73-REFERENCE-NAM-5.0. The aligned reads were calculated with FeatureCounts v1.6.2. HISAT v2.1.0 was used to construct the index and to compare the clean reads to the reference genome [28].

Afterward, StringTie v1.3.4d was used for the prediction of new genes [29]. featureCounts v1.6.2 and StringTie v1.3.4d were used to calculate the gene alignment and the FPKM. DESeq2 v1.22.1 and edgeR v3.24.3 were used to analyze the differentially expressed genes between the two groups [30–32], and the *p*-values were corrected using the Benjamini–Hochberg method. The corrected *p*-values and |log2foldchange| were used as the thresholds for significant difference expression. A hypergeometric distribution test was used for KEGG and GO term analyses [33–35].

2.5. RT-qPCR Analysis

A SteadyPure plant RNA extraction kit (AG) was used to extract the total RNA. MonScript™ RTIII ALL-in-One Mix with dsDNase (Monad) was used for reverse transcription. SYBR Green Pro Taq (AG) was used with an ABI 7500 Real-Time PCR System for fluorescence quantification; the amplified product was diluted to 500 ng/μL. The relative gene expression was calculated with the $^{\Delta\Delta}$Ct method. UBQ and AT1 were used as actins [36,37]. The primers used in this experiment are listed in Supplementary Table S1.

2.6. Statistical Analysis

Statistical analyses were performed using an ANOVA of Student's *t*-test ($p < 0.05$; LSD and Duncan test) in GraphPad Prism8 software (version 8.0.2, GraphPad Software, San Diego, CA, USA). All experiments were repeated at least three times. $p = 0.05$ indicated significant values.

3. Results

3.1. *zmcop1-1* and *zmcop1-2* Are Two Loss-of-Function Mutants

First, to understand the conservation of *COP1* in different species, we downloaded and analyzed *COP1* sequences. We found that *ZmCOP1* is highly similar to *AtCOP1* at the DNA sequence level (Figure S1). We then ordered two loss-of-function *zmcop1* mutants from the EMS mutant library (www.elabcaas.cn/memd/ (accessed on 14 September 2021)), naming them *zmcop1-1* and *zmcop1-2*, respectively. The two mutants were generated from B73 and self-fertilized for four generations. Through a Sanger sequencing analysis, we confirmed that both *zmcop1-1* and *zmcop1-2* were *zmcop1* stop-gain mutants that changed from TGA to TAA at the 1157th nucleotide and the 4608th nucleotide (Figure 1A,B), respectively. The mutation in *zmcop1-1* was located in the COIL helix domain; the base change led to the losses of the COIL and WD40 domains. The mutation in *zmcop1-2* was located between the COIL domain and the WD40 domain (Figure 1C), affecting the latter. It has been reported that WD40 is important in seedling and flower development as well as in light signal transmission and perception [38]. Thus, we propose that *zmcop1-1* and *zmcop1-2* may affect plant growth and development.

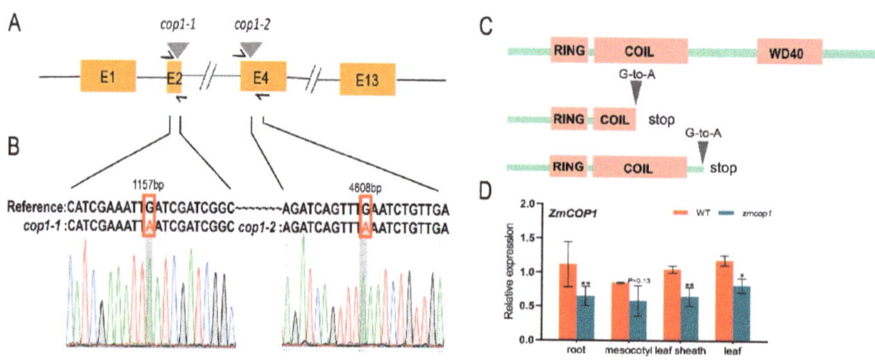

Figure 1. Genotyping expression patterns of *ZmCOP1*. (**A**) Illustration of the mutation sites in *zmcop1-1* and *zmcop1-2*. (**B**) Genotyping of *zmcop1-1* and *zmcop1-2*. B73 was used as a reference. As shown in the figure, G was mutated to A at the 1157th nucleotide in *zmcop1-1* and G was mutated to A at the 4608th nucleotide in *zmcop1-2*. (**C**) A model of the affected protein domains in *zmcop1-1* and *zmcop1-2*. (**D**) Expression patterns of *ZmCOP1* in different tissues (root, mesocotyl, leaf and leaf sheath). The maize seedlings were grown in darkness and at room temperature for 5 days. Data are means ± SD of at least three biological replicates; * $p < 0.05$, ** $p < 0.01$.

To understand the expression pattern of *ZmCOP1* in maize seedlings, we carried out an RT-qPCR analysis. We found that in 5-day-old etiolated seedlings in B73 and *zmcop1-1*, *ZmCOP1* was expressed in almost all tissues, indicating its role. The expression of *ZmCOP1* decreased significantly in *zmcop1-1* compared to in B73 by 42%, 40%, 38% and 31% in the root, mesocotyl, leaf sheath and leaf, respectively (Figure 1D).

3.2. *zmcop1-1* and *zmcop1-2* Shortened Mesocotyl Elongation

In order to study the function of *ZmCOP1* in mesocotyl elongation, we phenotypically analyzed the mesocotyl lengths of *zmcop1-1* and *zmcop1-2*, using B73 as a control (Figure 2A). When grown in the dark, B73 had a mesocotyl length of about 7.4 cm, while *zmcop1-1* and *zmcop1-2* each had a mesocotyl length of about 6.5 cm (Figure 2B). The mesocotyl length of *zmcop1-1* and *zmcop1-2* was significantly shorter than that of B73, with about a 12% reduction. The seedling length was 16.0 cm in the wild type, 12.1 cm in *zmcop1-1* and 14.6 cm in *zmcop1-2*. *zmcop1-1* and *zmcop1-2* showed significantly lower seedling lengths than the wild type (Figure 2C).

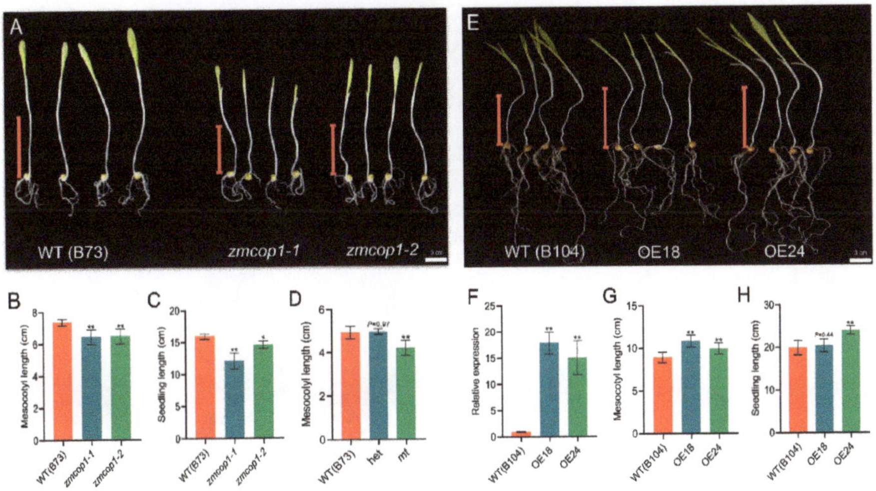

Figure 2. Phenotypic analyses of *zmcop1* mutants and overexpression transgenic lines. (**A**) Seedling phenotypes of wild-type B73, *zmcop1-1* and *zmcop1-2* grown in darkness for 7 days. Bars: 3 cm. (**B**) Quantification of mesocotyl length. (**C**) Quantification of seedling length. (**D**) Mesocotyl length segregation from a *zmcop1-1* heterozygous ear. (**E**) Seedling phenotypes of wild-type B104 and *ZmCOP1* overexpression lines (OE18, OE24, respectively) in darkness for 7 days. Bars: 3 cm. (**F**) Expressions of *ZmCOP1* in WT B104 (OE18, OE24), which was grown in darkness for 7 days. (**G**) Quantification of mesocotyl length. (**H**) Quantification of seedling length. Data are means ± SD of at least 10 biological replicates. Asterisks indicate significant differences in a two-way ANOVA (* $p < 0.05$, ** $p < 0.01$).

To understand whether the phenotype was caused by the *zmcop1* mutation, we performed a chi-square test using *zmcop1-1*. We found that in 121 seedlings, 100 showed the wild-type phenotype while 21 showed the short mesocotyl phenotype, which is inconsistent with the 3:1 segregation ratio (χ^2 = 3.160, df = 1). These data indicate that the short mesocotyl phenotype is caused by one gene.

We then genotyped the *zmcop1-1* seedlings and found that the length was 4.9 cm for the wild-type genotype, 5.0 cm for the heterozygous genotype and 4.2 cm for the mutated homozygous genotype. A statistical analysis showed that there were no differences between the wild-type and the heterozygous genotype (Figure 2D), indicating that the mutation genotype is recessive. This result is consistent with the phenotype of *cop1* in *Arabidopsis thaliana* grown in the dark, which has a short hypocotyl length and plant height [9].

3.3. *ZmCOP1* Overexpression Lines Showed Longer Mesocotyl Lengths

To understand whether overexpression of *ZmCOP1* promotes the length of etiolated seedling mesocotyls, we generated and planted transgenic lines while using B104 as a genetic background (Figure 2E). We gained more than 20 *ZmCOP1* OE lines and verified with RT-qPCR that the *ZmCOP1* in line 18 and line 24 was overexpressed more than tenfold compared with that in wild-type B104 (Figure 2F). We found that in darkness, the length of the etiolated seedling mesocotyl was about 10.8 cm in OE18, about 9.9 cm in OE24 and about 8.9 cm in wild-type B104 (Figure 2G). Thus, the mesocotyl lengths of the OEs were significantly longer than that of the wild type. We further measured the whole seedling length using these materials and found that the seedling length was 20.4 cm in OE18, 24.0 cm in OE24 and 19.8 cm in wild-type B104 (Figure 2H). The results showed that the high expression of *ZmCOP1* promoted the seedling hypocotyl length.

3.4. The Expressions of ZmHY5 and Other Light Genes Are Regulated by ZmCOP1

HY5 and HY5L have been reported to interact with COP1 to inhibit the morphogenesis of the hypocotyl elongation of *Arabidopsis* seedlings [9,22]. To understand whether *ZmHY5* is regulated by *ZmCOP1*, we detected the expressions of *ZmHY5* and *ZmHY5L* in the 5-day-old etiolated seedlings. We found that the *ZmHY5* and *ZmHY5L* in *zmcop1-1* decreased by 20% and 13%, respectively (Figure 3A,B), indicating that the *ZmCOP1* led to expression changes for *ZmHY5* and *ZmHY5L* in the etiolated maize seedlings. However, contrasting the fact that *AtHY5* is degraded by *AtCOP1* [39], it seems that *ZmHY5* is stabilized by *ZmCOP1*. This needs to be studied further.

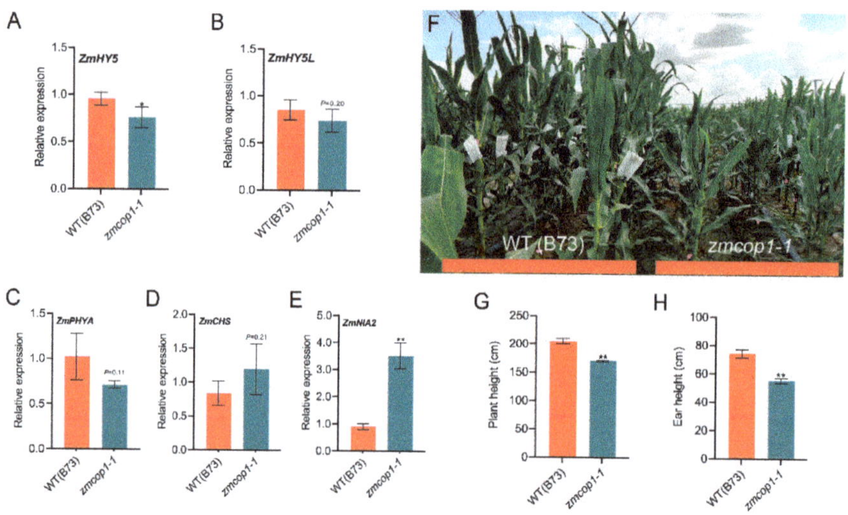

Figure 3. Gene expressions in and plant architecture of *zmcop1-1*. (**A**,**B**) Expressions of *ZmHY5* and *ZmHY5L* in wild-type B73 and *zmcop1* seedlings that were grown in darkness for 5 days. (**C**–**E**) Differential expressions of several light-regulating genes in wild-type B73 and *zmcop1* seedlings grown in darkness for 7 days. (**C**) Expressions of *ZmPHYA* in wild-type B73 and the *zmcop1-1* mutant. (**D**) Expressions of *ZmCHS* in wild-type B73 and the *zmcop1-1* mutant. (**E**) Expressions of *ZmNIA2* in wild-type B73 and the *zmcop1-1* mutant. (**F**) Plant architecture of *zmcop1-1* at silking time. (**G**) Quantification of plant height during the silking stage. (**H**) Quantification of ear height during the silking stage. Asterisks indicate significant differences between WT B73 and *zmcop1* using a Student's *t*-test and a two-way ANOVA (* $p < 0.05$, ** $p < 0.01$).

In order to understand whether *ZmCOP1* is involved in the absorption and utilization of light in maize, we identified the the expressions of several key light-regulating genes, such as *ZmPHYA*, *ZmCHS* and *ZmNIA2* [15]. We found that the expression of *ZmPHYA* in *zmcop1* was slightly but not statistically significantly lower than that in the WT (Figure 3C). The expressions of *ZmCHS* and *ZmNIA2* in *zmcop1* were higher than those in the WT, increased by 30% and 74%, respectively (Figure 3D,E). The phenomenon of *ZmCOP1* affecting the expression of photoregulatory factors indicated that the function of *ZmCOP1* is conserved for *AtCOP1*.

3.5. *ZmCOP1 Affects Plant Height*

In order to explore whether *ZmCOP1* affects plant development in the light, we carried out plant height phenotyping between WT B73 and *zmcop1* at the silking stage (Figure 3F). We found that the plant height was 204.8 cm in wild-type B73 and 170.4 cm in *zmcop1-1*. The plant height was lower by 17% in *zmcop1-1* than in the WT (Figure 3G). The ear height in the *zmcop1-1* mutant was also lower than that in the WT, reduced by 27% (Figure 3H). However, there was no significant difference in either the plant height or the ear height between the *zmcop1-2* mutant and wild-type B73; this needs to be studied further.

Chlorophyll is one of the most important photosynthetic pigments in plants, and its content directly affects the intensity of plant photosynthesis [40]. In order to clarify the effect of *ZmCOP1* on light absorption during maize's growth period, we determined the relative content of chlorophyll in its leaves. We found that there was little difference in the SPAD values of wild-type B73, *zmcop1-1* and *zmcop1-2* in the first and ear leaves (Figure S2).

3.6. *GO Analysis Showed That the DEGs Are Related to Hormone Signal Transduction*

To further explore the functional mechanism of *ZmCOP1*, we performed a RNA sequencing analysis using *zmcop1-1* and wild-type B73. We found 33 DEGs, of which 19 were up-regulated and 14 were down-regulated (Figure 4A). To verify the expression patterns of the DEGs, we performed RT-qPCR. We confirmed that the results thereof were consistent with those of the RNA-seq (Figure 4B). A volcano map was used to show the overall distributions of differential genes in the WT and *zmcop1* (Figure 4C). Based on the expression patterns between the WT and *zmcop1*, we clustered the differentially expressed genes into groups (Figure S3). The genes that showed consistent expression patterns in all of the *zmcop1* replicates were speculated to have similar functions.

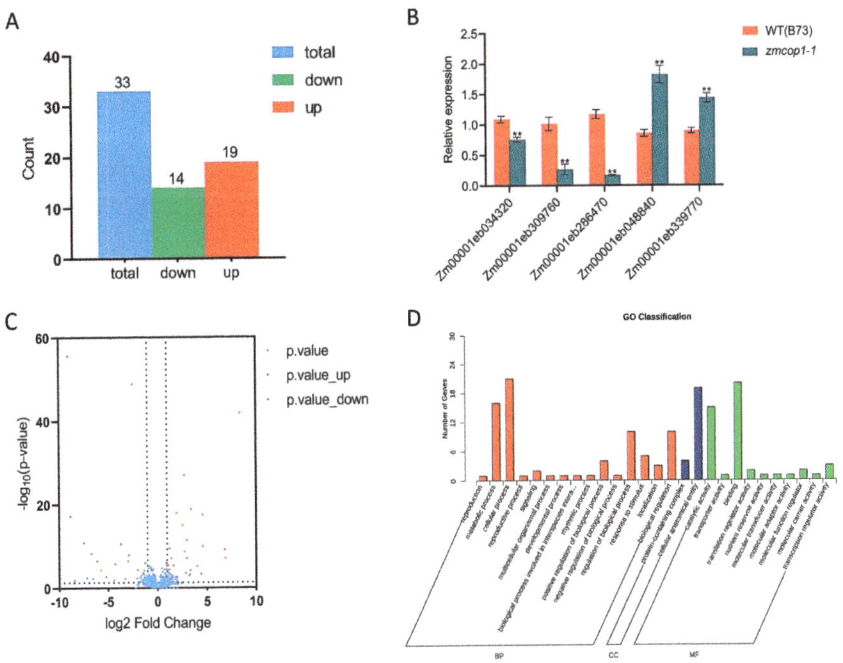

Figure 4. RNA-seq results showed that *zmcop1-1* mutation led to expression changes in some genes. (**A**) Differentially expressed genes between wild-type B73 and the *zmcop1-1* mutant. (**B**) Verification of

the DEGs with RT-qPCR. Asterisks indicate significant differences between WT B73 and *zmcop1* using a Student's *t*-test and a two-way ANOVA (** $p < 0.01$. (**C**) DEG volcano map. Red dots represent up-regulated differential genes in *zmcop1-1*, green dots represent down-regulated differential genes and blue dots represent non-differentially expressed genes. (**D**) GO bar chart. The abscissa denotes secondary GO entries, and the ordinate indicates the number of differential genes in the GO entries.

To understand the biological information of the DEGs, a GO (gene ontology) analysis was performed. According to the threshold of $p \leq 0.05$, the DEGs were divided into three main functional categories: BP (biological process), CC (cellular component) and MF (molecular function) (Figure 4D). The regulations of biological (10 DEGs), metabolic (16 DEGs) and cellular processes (21 DEGs) were enriched in the BP category. The cellular anatomical entity (19 DEGs) was enriched in the CC category. The catalytic activity (15 DEGs) and binding (20 DEGs) were enriched in the MF category. The GO analysis showed that the DEGs are related to plant hormone signal transduction and metabolite biosynthesis.

3.7. KEGG Showed That DEGs Are Related to Hormone Signal Transduction

To understand which biological processes DEGs participate in, KEGG (Kyoto Encyclopedia of Genes and Genomes) enrichment analyses were conducted (Figure 5A) [35]. We found that *ZmCOP1* regulates a wide range of KEGG pathways, such as those for MAPK hormone signal transduction, ribosome formation, RNA transport, glyoxylic acid and dicarboxylic acid metabolism and biosynthesis of secondary metabolites. These pathways are closely related to plant hormone production, genetic material changes and metabolite biosynthesis.

It has been reported that *ZmCOP1* regulates plant cell elongation and division by regulating the BR signal transduction pathway [41]. In *Arabidopsis thaliana*, both JA and BR play an important role in regulating cell and hypocotyl elongation [41–43]. We found that *Zm00001eb309760* was rarely expressed in *zmcop1* (Figure 5B). *Zm00001eb309760* is annotated to encode a leucine-rich repeat receptor-like protein kinase family protein. *Zm00001eb309760* may participate in the biosynthesis of BR. In addition, we found that a new gene, named *novel.1754*, was increased in the *zmcop1* mutant (Figure 5C). The KEGG analysis showed that *novel.1754* participates in the biosynthesis of jasmonic acid by regulating MYC2 and also regulates the α-linolenic acid metabolic pathway (Figure 5D). In summary, based on our KEGG results, we propose that *ZmCOP1* participates in the ETH, BR and JA plant hormone pathways.

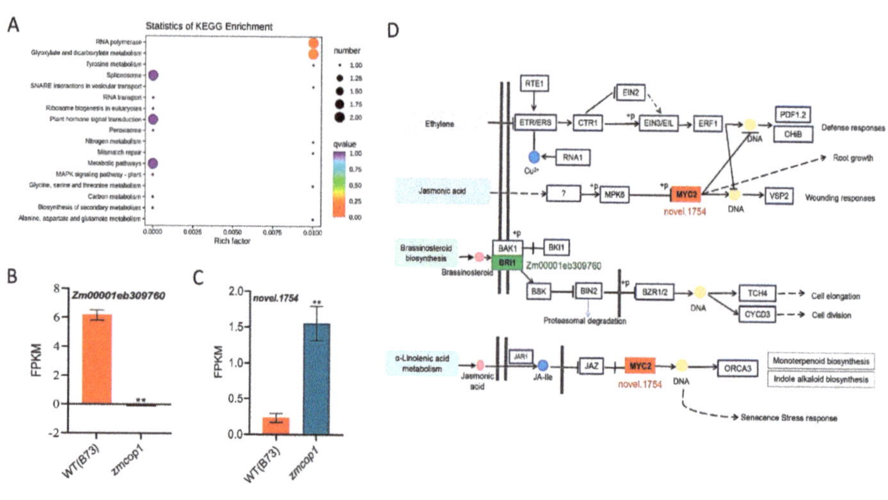

Figure 5. KEGG analysis of DEGs. (**A**) KEGG enrichment analysis scatter plot. The ordinate represents the KEGG path. The abscissa represents the rich factor. The larger the rich factor, the greater the degree

of enrichment. The larger the point, the larger the number of differential genes enriched in the pathway. The redder the color of the dot, the more significant the enrichment. (**B**) Expression of *Zm00001eb309760* in RNA-seq. Asterisks indicate significant differences between wild-type B73 and the mutant material using a Student's *t*-test (** $p < 0.01$). (**C**) Expression of *novle.1754* in RNA-seq. Asterisks indicate significant differences between wild-type B73 and the mutant material using a Student's *t*-test (** $p < 0.01$). (**D**) KEGG pathway maps. The substances marked in red boxes represent up-regulated genes, while those marked in green boxes represent down-regulated genes.

4. Discussion

A previous study showed that *ZmCOP1* could restore the phenotype of an *atcop1-4* mutant in *Arabidopsis thaliana*, including the gene expressions for hypocotyl length, cotyledon openings, chlorophyll levels and light response [9]. In our study, the mesocotyl of the *zmcop1* mutants was significantly shorter than that of the WT in maize seedlings grown in the dark for 7 days (Figure 2B). By measuring the height of the seedlings, we found that the seedling height of the *zmcop1* mutants was also shorter (Figure 2C). However, unlike with the *atcop1* phenotype in *Arabidopsis thaliana*, we did not observe significant changes in the cotyledons at the maize seedling stage. We speculate that the function of *COP1* in *Arabidopsis thaliana* and maize is conservative, but there are also some differences.

COP1-HY5 forms the core complex that controls plant photomorphogenesis [22,44]. COP1 and HY5 play antagonistic roles in response to light signals and in the regulation of seedling morphogenesis. HY5 is a key transcription factor involved in the inhibition of hypocotyl elongation. We found that the contents of *ZmHY5* and *ZmHY5L* were both decreased in the *zmcop1* mutants (Figure 3A,B). It has been reported that *AtHY5* is degraded by *AtCOP1* [39], and our results showed that *ZmHY5* is stabilized by *ZmCOP1* at the RNA level. Whether ZmHY5 is degraded by ZmCOP1 at the protein level needs to be studied further.

Low expression of *AtCOP1* has been found to lead to differential expression of normal light-regulated genes in dark-treated materials. *atcop1* mutation inhibits photomorphogenesis and the elongation of hypocotyl [45,46]. In our research, through RT-qPCR, we confirmed that the expression of the light-regulated gene *ZmPHYA* in the *zmcop1* mutant seedlings treated in darkness for 7 days was slightly lower than that in the WT (Figure 3C). We speculate that some light-regulating genes in maize *zmcop1* are also affected, and that mutation of *zmcop1* affects plant morphogenesis.

It has been reported that *nia1* and *nia2* blossom and mature earlier than the WT in *Arabidopsis thaliana*, indicating that *AtNIA1* and *AtNIA1* are involved in the regulation of plant flowering [47]. We observed that the tasseling and flowering times of the *zmcop1* mutants were 3–5 days later than those of B73. Additionally, the expression of *ZmNIA2* was increased in the *zmcop1* mutants compared to in B73 (Figure 3E). Therefore, we speculated that the late flowering phenotype was caused by the high expression of *ZmNIA2* in the *zmcop1* mutants.

Studies have shown that under low light intensity, HY5 in Arabidopsis is down-regulated by COP1-mediated ubiquitination and degradation. BIN2 is partially inactivated with repression of the transcriptional activity of HY5. In contrast, BZR1 is accumulated (Li et al., 2020). As a result, the hypocotyl length is promoted by the enhanced transcriptional activity of the related genes. With an increase in light intensity, the elongation of Arabidopsis hypocotyls would be inhibited. Our RNA-seq results showed down-regulation of BRI1 in *zmcop1* mutants (Figure 5B). Considering the function of *AtCOP1*, we propose that *zmcop1* mutation first affects the downstream BR synthesis pathway, inhibiting cell elongation and division, and affects plant height afterward (Figure 5D).

5. Conclusions

We experimentally demonstrated the roles of *ZmCOP1* in maize morphogenesis. We found that *ZmCOP1* affects the elongation of maize mesocotyl in darkness and affects plant height in the light. We confirmed that *ZmCOP1* regulates the expressions of several key light factors in the same way as *AtCOP1*. Finally, we identified some DEGs and showed that *ZmCOP1* may control maize morphogenesis by regulating genes involved in the plant phytohormone pathway. Our data can be applied in maize performance improvement.

Supplementary Materials: The following supporting information can be downloaded at: https://www.mdpi.com/article/10.3390/life13071522/s1, Figure S1: Phylogenetic analysis of ZmCOP1. (A) Homology analysis of COP1 proteins in maize, sorghum, rice, grape, Arabidopsis, soybean and poplar. (B) Phylogenetic tree analysis showed that ZmCOP1 and AtCOP1 proteins are highly similar; Figure S2: Measurement of relative content of chlorophyll. (A) The leaf site of relative content of chlorophyll. (B) SPAD of the ear leaves. (C) SPAD of the first leaves. Data are mean ± SD of at least 20 plants.; Figure S3: RNA-seq results showed that zmcop1 mutation led to changes in the expression of some genes. (A) Differential gene clustering heat map. Abscissa denotes sample name and hierarchical clustering result, ordinate indicates differential gene and hierarchical clustering result. Red indicates high expression and green indicates low expression. (B) Kmeans cluster diagram. The Abscissa represents the sample, and the ordinate represents the standardized expression.; Table S1: The primers used in this experiment.

Author Contributions: L.C. designed the experiments, performed the experiments and data analysis, wrote the original draft and wrote and revised the manuscript. Q.L. performed the experiments and data analysis. M.W. performed the experiments and data analysis and revised the manuscript. F.X. performed the experiments. K.L. performed the experiments. R.Y. performed the experiments. M.S. performed the experiments. H.Z. provided theoretical guidance. J.G. designed the experiments and performed the data analysis. J.C. designed the experiments, performed the data analysis and revised the manuscript. F.J. designed the experiment, performed the data analysis and wrote and revised the manuscript. All authors have read and agreed to the published version of the manuscript.

Funding: This work was supported by the Taishan Scholars Program of Shandong Province (tsqn201909134), the Maize Industry Technology System Genetic and Breeding Positions in Shandong (SDAIT-02-01), the Shandong Provincial Natural Science Foundation (ZR2021QC179 and ZR2022MC193), the National Key Research and Development Program of China (2021YFD1900903), Well-Breed Engineering of Shandong province (2021LZGC022), the science and technology commissioner of Tai'an (2021TPY007) and the Talent Introduction Special Funds of Qingdao Agricultural University (665/1119016 and 665/1120014).

Institutional Review Board Statement: Not applicable.

Informed Consent Statement: Not applicable.

Data Availability Statement: Data will be made available on request.

Acknowledgments: We thank the members of the Laboratory of Maize Molecular Genetics and Breeding for their assistance in this study. The mutant materials used in this study were provided by Chunyi Zhang from the Institute of Biotechnology, Chinese Academy of Agricultural Sciences and Xiaoduo Lu from Qilu Normal University.

Conflicts of Interest: The authors declare that they have no known competing financial interest or personal relationship that could have appeared to influence the work reported in this paper.

References

1. Liu, X.; Qin, T.; Ma, Q.; Sun, J.; Liu, Z.; Yuan, M. Light-regulated hypocotyl elongation involves proteasome-dependent degradation of the microtubule regulatory protein WDL3 in *Arabidopsis*. *Plant Cell* **2013**, *255*, 1740–1755. [CrossRef] [PubMed]
2. Brandizzi, F. Divide, expand, differentiate—New insights on plant organ growth through cytokinin signaling. *Plant J.* **2019**, *97*, 803–804. [CrossRef] [PubMed]
3. Nakano, T. Hypocotyl Elongation: A Molecular Mechanism for the First Event in Plant Growth That Influences Its Physiology. *Plant Cell Physiol.* **2019**, *60*, 933–934. [CrossRef] [PubMed]
4. Li, J.; Terzaghi, W.; Gong, Y.; Li, C.; Ling, J.J.; Fan, Y. Modulation of BIN2 kinase activity by HY5 controls hypocotyl elongation in the light. *Nat. Commun.* **2020**, *111*, 1592. [CrossRef]

5. Liu, H.; Lin, R.; Deng, X.W. Photobiology: Light signal transduction and photomorphogenesis. *J. Integr. Plant Biol.* **2020**, *62*, 1267–1269. [CrossRef]
6. Legris, M.; Ince, Y.C.; Fankhauser, C. Molecular mechanisms underlying phytochrome-controlled morphogenesis in plants. *Nat. Commun.* **2019**, *10*, 5219. [CrossRef]
7. Jing, Y.; Lin, R. Transcriptional regulatory network of the light signaling pathways. *New Phytol.* **2020**, *227*, 683–697. [CrossRef]
8. Roeber, V.M.; Bajaj, I.; Rohde, M.; Schmülling, T.; Cortleven, A. Light acts as a stressor and influences abiotic and biotic stress responses in plants. *Plant Cell Environ.* **2021**, *44*, 645–664. [CrossRef]
9. Huai, J.; Jing, Y.; Lin, R. Functional analysis of *ZmCOP1* and *ZmHY5* reveals conserved light signaling mechanism in maize and Arabidopsis. *Physiol. Plant.* **2020**, *169*, 369–379. [CrossRef]
10. Hoecker, U.; Quail, P.H. The Phytochrome A-specific Signaling Intermediate SPA1 Interacts Directly with COP1, a Constitutive Repressor of Light Signaling in *Arabidopsis*. *J. Biol. Chem.* **2001**, *276*, 38173–38178. [CrossRef]
11. Jang, I.-C.; Henriques, R.; Seo, H.S.; Nagatani, A.; Chua, N.-H. *Arabidopsis* Phytochrome Interacting Factor Proteins Promote Phytochrome B Polyubiquitination by COP1 E3 Ligase in the Nucleus. *Plant Cell* **2010**, *22*, 2370–2383. [CrossRef]
12. Xu, D.; Lin, F.; Jiang, Y.; Huang, X.; Li, J.; Ling, J. The RING-Finger E3 Ubiquitin Ligase COP1 SUPPRESSOR1 Negatively Regulates COP1 Abundance in Maintaining COP1 Homeostasis in Dark-Grown Arabidopsis Seedlings. *Plant Cell* **2014**, *265*, 1981–1991. [CrossRef]
13. Pham, V.N.; Kathare, P.K.; Huq, E. Phytochromes and Phytochrome Interacting Factors. *Plant Physiol.* **2018**, *176*, 1025–1038. [CrossRef] [PubMed]
14. Paik, I.; Chen, F.; Pham, V.N.; Zhu, L.; Kim, J.I.; Huq, E. A phyB-PIF1-SPA1 kinase regulatory complex promotes photomorphogenesis in *Arabidopsis*. *Nat Commun.* **2019**, *101*, 4216. [CrossRef] [PubMed]
15. Deng, Y.; Wang, J.; Zhang, Z.; Wu, Y. Transactivation of *Sus1* and *Sus2* by Opaque2 is an essential supplement to sucrose synthase-mediated endosperm filling in maize. *Plant Biotechnol. J.* **2020**, *18*, 1897–1907. [CrossRef] [PubMed]
16. Ohgishi, M.; Saji, K.; Okada, K.; Sakai, T. Functional analysis of each blue light receptor, cry1, cry2, phot1, and phot2, by using combinatorial multiple mutants in Arabidopsis. *Proc. Natl. Acad. Sci. USA* **2004**, *101*, 2223–2228. [CrossRef]
17. Yu, X.; Liu, H.; Klejnot, J.; Lin, C. The Cryptochrome Blue Light Receptors. *Arab. Book* **2010**, *8*, e0135. [CrossRef]
18. Ponnu, J.; Riedel, T.; Penner, E.; Schrader, A.; Hoecker, U. Cryptochrome 2 competes with COP1 substrates to repress COP1 ubiquitin ligase activity during Arabidopsis photomorphogenesis. *Proc. Natl. Acad. Sci. USA* **2019**, *116*, 27133–27141. [CrossRef]
19. Wang, W.; Paik, I.; Kim, J.; Hou, X.; Sung, S.; Huq, E. Direct phosphorylation of HY5 by SPA kinases to regulate photomorphogenesis in Arabidopsis. *New Phytol.* **2021**, *230*, 2311–2326. [CrossRef]
20. Carranco, R.; Prieto-Dapena, P.; Almoguera, C.; Jordano, J. A seed-specific transcription factor, HSFA9, anticipates UV-B light responses by mimicking the activation of the UV-B receptor in tobacco. *Plant J.* **2022**, *1115*, 1439–1452. [CrossRef]
21. Zhu, L.; Bu, Q.; Xu, X.; Paik, I.; Huang, X.; Hoecker, U.; Deng, X.W.; Huq, E. CUL4 forms an E3 ligase with COP1 and SPA to promote light-induced degradation of PIF1. *Nat. Commun.* **2015**, *6*, 7245. [CrossRef]
22. Bhatnagar, A.; Singh, S.; Khurana, J.P.; Burman, N. HY5-COP1: The central module of light signaling pathway. *J. Plant Biochem. Biotechnol.* **2020**, *29*, 590–610. [CrossRef]
23. Lian, N.; Liu, X.; Wang, X.; Zhou, Y.; Li, H.; Li, J. COP1 mediates dark-specific degradation of microtubule-associated protein WDL3 in regulating Arabidopsis hypocotyl elongation. *Proc. Natl. Acad. Sci. USA* **2017**, *11446*, 12321–12326. [CrossRef]
24. Zheng, Y.; Cui, X.; Su, L.; Fang, S.; Chu, J.; Gong, Q. Jasmonate inhibits COP1 activity to suppress hypocotyl elongation and promote cotyledon opening in etiolated *Arabidopsis* seedlings. *Plant J.* **2017**, *906*, 1144–1155. [CrossRef]
25. Cao, J.; Liang, Y.; Yan, T.; Wang, X.; Zhou, H.; Chen, C. The photomorphogenic repressors BBX28 and BBX29 integrate light and brassinosteroid signaling to inhibit seedling development in Arabidopsis. *Plant Cell* **2022**, *346*, 2266–2285. [CrossRef]
26. Chen, Y.-F.; Etheridge, N.; Schaller, G.E. Ethylene Signal Transduction. *Ann. Bot.* **2005**, *95*, 901–915. [CrossRef]
27. Shi, H.; Liu, R.; Xue, C.; Shen, X.; Wei, N.; Deng, X.W. Seedlings Transduce the Depth and Mechanical Pressure of Covering Soil Using COP1 and Ethylene to Regulate EBF1/EBF2 for Soil Emergence. *Curr Biol.* **2016**, *262*, 139–149. [CrossRef]
28. Kim, D.; Langmead, B.; Salzberg, S.L. HISAT: A fast spliced aligner with low memory requirements. *Nat. Methods* **2015**, *12*, 357–360. [CrossRef]
29. Pertea, M.; Pertea, G.M.; Antonescu, C.M.; Chang, T.-C.; Mendell, J.T.; Salzberg, S.L. StringTie enables improved reconstruction of a transcriptome from RNA-seq reads. *Nat. Biotechnol.* **2015**, *33*, 290–295. [CrossRef]
30. Robinson, M.D.; McCarthy, D.J.; Smyth, G.K. EdgeR: A Bioconductor package for differential expression analysis of digital gene expression data. *Bioinformatics* **2010**, *26*, 139–140. [CrossRef]
31. Love, M.I.; Huber, W.; Anders, S. Moderated estimation of fold change and dispersion for RNA-seq data with DESeq2. *Genome Biol.* **2014**, *1512*, 550. [CrossRef]
32. Varet, H.; Brillet-Gueguen, L.; Coppee, J.Y.; Dillies, M.A. SARTools: A DESeq2- and EdgeR-Based R Pipeline for Comprehensive Differential Analysis of RNA-Seq Data. *PLoS ONE* **2016**, *116*, e0157022. [CrossRef] [PubMed]
33. Ashburner, M.; Ball, C.A.; Blake, J.A.; Botstein, D.; Butler, H.; Cherry, J.M.; Davis, A.P.; Dolinski, K.; Dwight, S.S.; Eppig, J.T.; et al. Gene ontology: Tool for the unification of biology. The Gene Ontology Consortium. *Nat. Genet.* **2000**, *25*, 25–29. [CrossRef]
34. Subramanian, A.; Tamayo, P.; Mootha, V.K.; Mukherjee, S.; Ebert, B.L.; Gillette, M.A.; Mesirov, J.P. Gene set enrichment analysis: A knowledge-based approach for interpreting genome-wide expression profiles. *Proc. Natl. Acad. Sci. USA* **2005**, *102*, 15545–15550. [CrossRef] [PubMed]

35. Kanehisa, M.; Araki, M.; Goto, S.; Hattori, M.; Hirakawa, M.; Itoh, M.; Katayama, T.; Kawashima, S.; Okuda, S.; Tokimatsu, T.; et al. KEGG for linking genomes to life and the environment. *Nucleic Acids Res.* **2008**, *36*, D480–D484. [CrossRef]
36. Deng, X.W.; Caspar, T.; Quail, P.H. copl: A regulatory locus involved in light-controlled development and gene expression in Arabidopsis. *Genes Dev.* **1991**, *5*, 1172–1182. [CrossRef] [PubMed]
37. Wang, F.; Yu, Z.; Zhang, M.; Wang, M.; Lu, X.; Liu, X. ZmTE1 promotes plant height by regulating intercalary meristem formation and internode cell elongation in maize. *Plant Biotechnol. J.* **2022**, *203*, 526–537. [CrossRef] [PubMed]
38. Mishra, A.K.; Puranik, S.; Prasad, M. Structure and regulatory networks of WD40 protein in plants. *J. Plant Biochem. Biotechnol.* **2012**, *21*, 32–39. [CrossRef]
39. Burko, Y.; Seluzicki, A.; Zander, M.; Pedmale, U.V.; Ecker, J.R.; Chory, J. Chimeric Activators and Repressors Define HY5 Activity and Reveal a Light-Regulated Feedback Mechanism. *Plant Cell* **2020**, *324*, 967–983. [CrossRef]
40. Zhou, J.; Su, X.; Zheng, S.; Wu, C.; Su, Y.; Jiang, Z.; Li, L.; Chen, S.; He, X. The *Arabidopsis* NuA4 histone acetyltransferase complex is required for chlorophyll biosynthesis and photosynthesis. *J. Integr. Plant Biol.* **2022**, *64*, 901–914. [CrossRef]
41. Nieto, C.; Luengo, L.M.; Prat, S. Regulation of COP1 Function by Brassinosteroid Signaling. *Front. Plant Sci.* **2020**, *11*, 1151. [CrossRef] [PubMed]
42. Tian, H.; Lv, B.; Ding, T.; Bai, M.; Ding, Z. Auxin-BR Interaction Regulates Plant Growth and Development. *Front. Plant Sci.* **2017**, *8*, 2256. [CrossRef] [PubMed]
43. Yi, R.; Yan, J.; Xie, D. Light promotes jasmonate biosynthesis to regulate photomorphogenesis in Arabidopsis. *Sci. China Life Sci.* **2020**, *63*, 943–952. [CrossRef] [PubMed]
44. Yamamoto, Y.Y.; Matsui, M.; Ang, L.H.; Deng, X.W. Role of a COP1 Interactive Protein in Mediating Light Regulated Gene Expression in *Arabidopsis*. *Plant Physiol.* **1998**, *10*, 1083–1094.
45. Lian, H.L.; He, S.B.; Zhang, Y.C.; Zhu, D.M.; Zhang, J.Y.; Jia, K.P. Blue-light-dependent interaction of crypto-chrome 1 with SPA1 defines a dynamic signaling mechanism. *Genes Dev.* **2011**, *2510*, 1023–1028. [CrossRef] [PubMed]
46. Sheerin, D.J.; Menon, C.; Oven-Krockhaus, S.z.; Enderle, B.; Zhu, L.; Johnen, P. Light-activated phytochrome A and B interact with members of the SPA family to promote photomorphogenesis in *Arabidopsis* by reorganizing the COP1/SPA complex. *Plant Cell* **2015**, *271*, 189–201. [CrossRef]
47. Seligman, K.; Saviani, E.E.; Oliveira, H.C.; Pinto-Maglio, C.A.F.; Salgado, I. Floral Transition and Nitric Oxide Emission during Flower Development in *Arabidopsis thaliana* is Affected in Nitrate Reductase-Deficient Plants. *Plant Cell Physiol.* **2008**, *49*, 1112–1121. [CrossRef]

life

Review

The Universally Conserved Unconventional G Protein YchF Is Critical for Growth and Stress Response

Zhaoheng Lin †, Rongfang Li †, Zhiwei Han, Yi Liu, Liyang Gao, Suchang Huang, Ying Miao and Rui Miao *

Fujian Provincial Key Laboratory of Plant Functional Biology, College of Life Sciences, Fujian Agriculture and Forestry University, Fuzhou 350002, China
* Correspondence: ruimiao@fafu.edu.cn; Tel.: +86-591-8373-7535
† These authors contributed equally to this work.

Abstract: The ancient guanine nucleotide-binding (G) proteins are a group of critical regulatory and signal transduction proteins, widely involved in diverse cellular processes of all kingdoms of life. YchF is a kind of universally conserved novel unconventional G protein that appears to be crucial for growth and stress response in eukaryotes and bacteria. YchF is able to bind and hydrolyze both adenine nucleoside triphosphate (ATP) and guanosine nucleoside triphosphate (GTP), unlike other members of the P-loop GTPases. Hence, it can transduce signals and mediate multiple biological functions by using either ATP or GTP. YchF is not only a nucleotide-dependent translational factor associated with the ribosomal particles and proteasomal subunits, potentially bridging protein biosynthesis and degradation, but also sensitive to reactive oxygen species (ROS), probably recruiting many partner proteins in response to environmental stress. In this review, we summarize the latest insights into how YchF is associated with protein translation and ubiquitin-dependent protein degradation to regulate growth and maintain proteostasis under stress conditions.

Keywords: YchF; growth; stress response; P-loop NTPase; ribosome; proteosome; protein translation; protein degradation

Citation: Lin, Z.; Li, R.; Han, Z.; Liu, Y.; Gao, L.; Huang, S.; Miao, Y.; Miao, R. The Universally Conserved Unconventional G Protein YchF Is Critical for Growth and Stress Response. *Life* **2023**, *13*, 1058. https://doi.org/10.3390/life13041058

Academic Editors: Sen Meng and Jie Luo

Received: 15 February 2023
Revised: 17 April 2023
Accepted: 18 April 2023
Published: 20 April 2023

1. Introduction

The ancient guanine nucleotide-binding (G) proteins, namely phosphate-binding-loop guanosine triphosphatases (P-loop GTPases), play crucial roles in protein synthesis and cellular signaling transduction among all kingdoms of life [1]. G proteins possess an active guanosine triphosphate (GTP) bound state and inactive guanosine diphosphate (GDP) bound state in a cyclic manner through loading GTP and hydrolyzing GTP to GDP [2,3]. G proteins receive upstream environmental signaling and transduce to downstream effectors. G protein molecular switches between "on" and "off" are coordinated by three G-protein-regulator families, including GTPase-activating proteins (GAPs), guanine nucleotide exchange factors (GEFs), and guanosine nucleotide dissociation inhibitors (GDIs). Nucleotide hydrolysis is accelerated by GAPs with the results of signal termination, while GEFs relieve GDP and replace GTP with G proteins, thereby activating the G proteins and turning on the signal transduction [1,4,5]. In contrast to GEFs, GDIs prevent the exchange of GTP with GDP and maintain G proteins in an inactive GDP-bound state, but the inhibition is revertible by the GEFs depending on the environmental stimuli [2,6,7].

On the basis of sequence and structural features, G proteins can be divided into two large distinct superclasses: The translation Factors (TRAFAC) superclass and Signal Recognition GTPases and the MinD and BioD (SIMIBI) superclass [8]. Secondly, signal-transducing G proteins comprise heterotrimeric G proteins, small G proteins, and many unconventional G proteins [4,9]. First, heterotrimeric G proteins are composed of G protein α subunits (Gα), Gβ and Gγ subunits. The GTPase (G) domain of heterotrimeric G protein is in the Gα subunit. The seven-transmembrane-spanning (7TM) G protein-coupled receptors (GPCR) work as GEFs to catalyze the exchange of GDP for GTP on the Gα subunit, while regulator of G-protein signaling (RGS) stimulates Gα subunit GTPase activity as GAPs [10,11]. In the presence of GDP, the GDP-bound Gα subunit integrates with the Gβγ heterodimer as a ternary complex intracellularly anchoring to GPCR at the plasma membrane. In the presence of GTP, however, the GTP-bound Gα subunit goes through a conformational change that allows heterotrimeric G protein dissociation into the Gα subunit and Gβγ heterodimer. Then, the Gα subunit and Gβγ heterodimer couple to their own effectors for signaling transduction [12]. Secondly, small G proteins hold a ~170 amino acid residue core G domain along with extra N-terminal and C-terminal extensions and can be divided into five families, such as Ras (Rat sarcoma), Rho (Ras homology), Arf (ADP-ribosylation factor), Rab (Ras-like in the brain), and Ran (Ras-like nuclear) [13]. Finally, heterotrimeric G proteins and small G proteins belong to the extended Ras-like family in the TRAFAC superclass, whereas the unique, unconventional G protein YchF subfamily is a part of the Obg family in TRAFAC superclass of G proteins [8,14].

2. Structure of G Domain among G Proteins

2.1. Structural Characterization of G Domain of G Proteins

All G proteins utilize the G domain to bind and hydrolyze nucleotides, which contains five structurally conserved motifs (G boxes): G1 motif (G1 box) adopting the sequence pattern GxxxxGK(S/T), G2 motif (G2 box) adopting the sequence pattern x(T/S)x, G3 motif (G3 box) adopting the sequence pattern hhhDxxG, G4 motif (G4 box) adopting the sequence pattern (N/T)KxD and G5 motif (G5 box) adopting the sequence pattern (T/G)(C/S)A [5] (Figure 1A). The so-called P-loop or walk A motif is the G1 box that binds to α- and β-phosphate of nucleotides. The walk B motif consists of a G2 box and a G3 box that anchor to the terminal γ-phosphate of nucleotide. The G3 box has a conserved aspartic acid (Asp/D) residue in contact with the co-factor magnesium (Mg^{2+}), which is crucial for nucleotide binding and hydrolysis (Figure 1A). In addition, the walk B motif overlapping with the switch I and switch II regions undergo a conformational change accompanied by nucleotide hydrolysis, which governs effector binding. The G4 box determines the guanosine or adenosine signature, and the G5 box supports specific recognition.

2.2. Structural Characterization of G Domain of YchF

The G domain of universally conserved unconventional G protein YchF maintains five fingerprint motifs as other G proteins. The G1, G2, G3, and G5 boxes are invariant with other G proteins, but the G4 box in the YchF subfamily shows a nontypical (N/T)(M/L/V)xE amino acid sequence instead of (N/T)KxD (Figure 1A,B; Table 1). Thus, the members in the YchF subfamily are capable of binding and hydrolyzing both adenine nucleoside triphosphate (ATP) and guanosine nucleoside triphosphate (GTP) [9].

Table 1. Summary of function-related amino acid residues in YchF subfamily.

Homolog	Species	Residues	Location	Supportive Reasons/Effects	Functions	References
E. coli YchF	*Escherichia coli*	His114	A highly flexible loop of G domain	Supporting the flexible loop to reach a catalytically active conformation	Critical for ATPase activity (+)	[15]
E. coli YchF	*Escherichia coli*	Cys35	G2 motif	Allows YchF dimerization via a disulfide bridge	Critical for ATPase activity (−)	[16]
E. coli YchF	*Escherichia coli*	Lys78 (Arg)	G domain	YchF-K78A mutant shows similar hydrolysis activities in presence of Na^+ or K^+, but K78R mutant retained potassium specific stimulation of ATPase activity	Plays a key role in determining the potassium dependent ATPase activity	[17]
hOLA1	*Homo sapiens*	Leu96	G domain (next to G3 motif)	A conserved Gln residue involved in GTP hydrolysis in Ras-like GTPases has been replaced	Inactivates Ras-like GTPases	[14]
E. coli YchF	*Escherichia coli*	Ser16 (Ser36 in *H. sapiens*)	G1 motif	Ser16 phosphorylated when H_2O_2 absence; Dissociation of KatG	Supports the ATPase activity; Detoxifies H_2O_2	[14,18]
E. coli YchF	*Escherichia coli*	Leu76	G3 motif	Hallmark for HAS-NTPase	Slightly affects ATPase activity (+)	[15]
hOLA1	*Homo sapiens*	Thr37	G domain	The main chain amide of Thr37 contacts the α-phosphate of AMPPCP	Supports the ATPase activity	[14]
hOLA1	*Homo sapiens*	Ser36/Val33	G1 motif	The main chain amide of Ser36 and Val33 contacts the β-phosphate of AMPPCP	Supports the ATPase activity	[14]
hOLA1	*Homo sapiens*	Asn32	G1 motif	The main chain amide of Asn32 forms a hydrogen bond to the γ-phosphate of AMPPCP	Supports the ATPase activity	[14]
hOLA1	*Homo sapiens*	Asn230	G4 motif	Its mutation to alanine abolished nucleotide binding	Contribute to nucleotide binding	[14]
hOLA1	*Homo sapiens*	Leu231	G4 motif	Specificity for adenine binding is based on the interaction between the adenine N-6 group and Leu231 main chain CO in G4 motif	Make YchF preference for ATP rather than GTP	[14]
hOLA1	*Homo sapiens*	Ser310	TGS domain	The H-bond between Ser310 O_γ and the exocyclic N-6 of an adenine is formed in a position similar to the ppGpp O-6	Make YchF preference for ATP rather than GTP	[14]
hOLA1	*Homo sapiens*	Phe127	Coiled-coil domain	Mutating this residue to Ala diminishes ATP binding drastically	Contribute to base recognition	[14]
AtYchF1	*Arabidopsis thaliana*	Glu345	TGS domain	Conserved and solvent-exposed	Most critical for its interaction with the regulator, GAP1	[19]

Amino acids in brackets indicate that there are other known residues presenting at the same position in the orthologs; In the "functions" column, (+) means upregulation, and (−) means down-regulation.

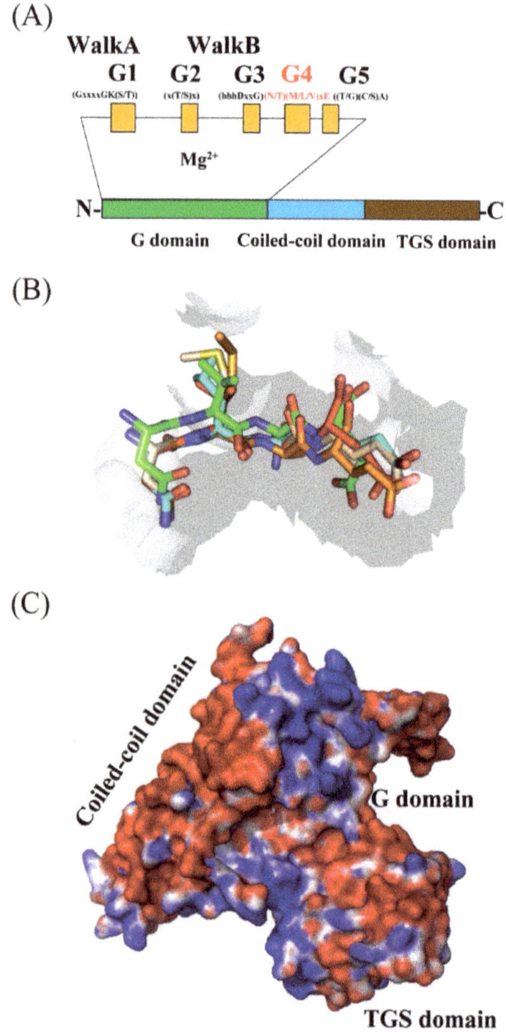

Figure 1. Structural characterization of the evolutionarily conserved unconventional G protein YchF. (**A**) Schematic representation of the structure of universally conserved unconventional G protein YchF. (**B**) Structural alignment of the nontypical G4 motif (N/T)(M/L/V)xE in the YchF subfamily (*E. coli* YchF is green, hOLA1 is cyan, *S. pombe* YchF is yellow, OsYchF1 is brown, *T. thermophilus* YchF is red). The amino acid residues are shown as sticks. (**C**) Electron density surface of the apo-structure of OsYchF1 (Protein Data Bank (PDB) code: 5EE0). Negatively charged amino acid residues are red, and positively charged amino acid residues are blue.

2.3. *Structural Comparison of G Domains among Selected YchF, Small G Protein, and Heterotrimeric G Protein α-Subunit*

Herein, a heterotrimeric G protein α-subunit in the rat (*Rattus norvegicus*) and a human (*Homo Sapien*) small G protein Ras-related G protein C was chosen to compare with OsYchF1, a rice (*Oryza sativa*) ortholog of YchF in plants. In contrast to *R. norvegicus* heterotrimeric G protein α-subunit and human Ras-related G protein C, the novel G4 motif and G5 motif of OsYchF1 support either ATP or GTP binding in the nucleotide-binding site of OsYchF1 (Figures 1B and 2A). According to the crystal structure of OsYchF1 in the presence of the ATP non-hydrolyzed homolog AMPPNP (Protein Data Bank (PDB) code: 5EE3), the backbone carboxyl group of methionine (M231) in the G4 motif of OsYchF1 forms a hydrogen bond with the adenine 6-amino group of AMPPNP (Figure 2A) [9,20]. This allows for the non-hydrolytic AMPPNP to be able to fit into the OsYchF1 nucleotide-binding site (Figure 2A). The structural alignments of OsYchF1 with *R. norvegicus* heterotrimeric G protein α-subunit (PDB code: 1SVS) and human Ras-related G protein C (PDB code: 3LLU) revealed that the side chain of asparagine (Asn) in the G4 motif of OsYchF1 could not turn back and interact with the 2-amino group of guanosine, unlike the other two proteins. However, the crystal structure of OsYchF1 in the presence of GppNHp (PDB code: 5EE9), a non-hydrolyzed homolog of GTP, showed that the G5 motif of OsYchF1 can form a hydrogen bond with the guanosine base group of GppNHp. This finding partially explains why OsYchF1 is capable of binding to GTP as well (Figure 2B) [9,20]. Moreover, the G1 motif (P-loop) is highly conserved and consistent among OsYchF1, *R. norvegicus* heterotrimeric G protein α-subunit, and human Ras-related G protein C (Figure 2A,B). The G1 motif of OsYchF1 interacts with the triphosphate of nucleotides, resembling the G1 motif of *R. norvegicus* heterotrimeric G protein α-subunit and human Ras-related Protein C (Figure 2A,B). In the OsYchF1 G4 motif mutant, however, methionine (Met) and glutamine (Glu) were replaced by lysine (Lys) and aspartic acid (Asp), respectively. With this change in amino acids, OsYchF1 obtained GTP priority again, indicating that the OsYchF1 G4 motif indeed determines ATP or GTP recognition [9,20].

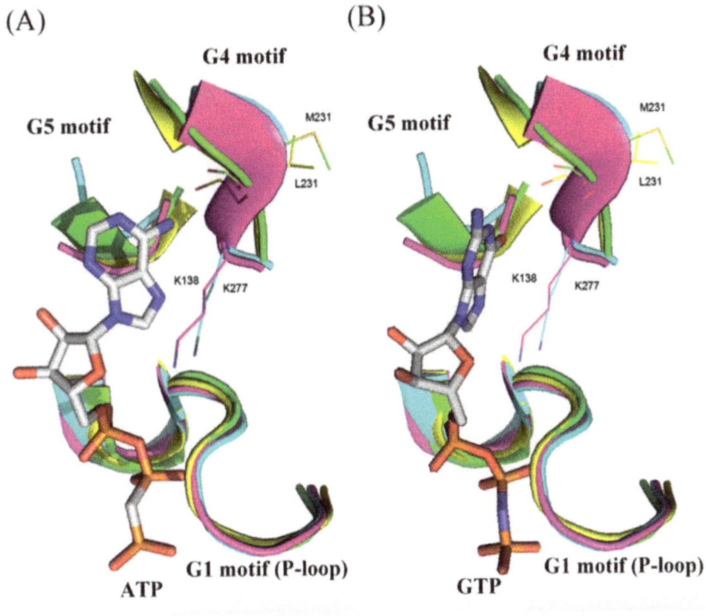

Figure 2. Structural alignments of OsYchF1 hOLA1 *H. sapiens* Ras-related G protein C and *R. norvegicus* heterotrimeric G protein α-subunit nucleotide-binding sites in the complex with nucleotides. (**A**) Structural alignments of OsYchF1 (PDB code: 5EE3), hOLA1 (PDB code: 2OHF), human Ras-related G protein C (HsRas C) (PDB code: 3LLU), and *R. norvegicus* heterotrimeric G protein α-subunit (RnHetero) (PDB code: 1SVS) nucleotide-binding site in the complex with the ATP non-hydrolyzed homolog AMPPNP. (**B**) Structural alignments of OsYchF1 (PDB code: 5EE9), hOLA1, HsRas C, and RnHetero nucleotide-binding site in the complex with the GTP non-hydrolyzed homolog GppNHp. AMPPNP, GppNHp, M-231, L-231, K-138, and K-277 are shown as sticks. The G1 motif (P-loop), G4 motif, and G5 motif are shown as cartoons (OsYchF1 is green, hOLA1 is yellow, human Ras-related G protein C is cyan, and *R. norvegicus* heterotrimeric G protein α-subunit is pink).

3. YchF Is Critical for Growth and Stress Response

3.1. YchF Works as a Conserved Negative Regulator in Response to Oxidative Stress

YchF is a universally conserved unconventional G-proteins in most organisms except archaea and consists of an N-terminal core G domain, inserted large coiled-coil domain, and C-terminal TGS (ThrRS, GTPase, and SpoT) domain potentially favoring the RNA binding and ubiquitin-dependent protein degradation (Figure 1A,C). Thus far, the available results suggest that YchF is probably a nucleotide-dependent translational factor associated with the ribosome and proteasome and likely links with other partner proteins as a unique negative regulator of the oxidative stress response (Table 2).

Table 2. Summary of main cellular elements interacting with the members of the YchF subfamily.

Homolog	Organism	Interactive Factors	Effects	References
OsYchF1	*Oryza sativa*	OsGAP1	Activating OsYchF1 GTPase and ATPase activity	[2,19,21,22]
AtYchF1	*Arabidopsis thaliana*	AtGAP1	Activating AtYchF1 GTPase and ATPase activity	[2,19,21,22]
AtYchF1	*Arabidopsis thaliana*	ppGpp	*AtYchF1* might be a critical regulator in controlling the cytosolic ppGpp-mediated growth inhibition in plants	[23]
E. coli YchF	*Escherichia coli*	30S ribosome, 70S ribosome	The 70S ribosome act as an ATPase activating factor (AAF) to stimulate YchF's ATPase activity	[17,24]
E. coli YchF	*Escherichia coli*	tRNA	YchF interacts with the 3′-CCA end of tRNA through its TGS-domain, indicating that YchF is involved in protein synthesis	[25]
T. cruzi YchF	*Trypanosoma cruzi*	26S Proteasome	*T. cruzi* YchF co-immunoprecipitates with a regulatory subunit of the *T. cruzi* proteasome, involving in protein degradation	[26,27]
E. coli YchF	*Escherichia coli*	KatG	YchF interacts with KatG and inhibit its catalase activity, revealing that YchF regulates the oxidative stress response	[16]
S. cerevisiae YchF	*Saccharomyces cerevisiae*	Eukaryotic translation elongation factor 1 (eEF1)	Supporting a role for YchF during translation	[28,29]
E. coli YchF	*Escherichia coli*	Translation initiation factor 3 (IF3)	YchF enhances the anti-association activity of IF3, stimulates the translation of leaderless mRNAs	[30]
E. coli YchF	*Escherichia coli*	Thioredoxin 1 (TrxA)	YchF dimer is dissociated by TrxA, which stimulates the ATPase activity	[16]
hOLA1	*Homo sapiens*	Eukaryotic elongation initiation factor 2 (eIF2)	hOLA1 effectively blocks the formation of TC (ternary complex) through its intrinsic GTPase activity, leading eIF2 unable to deliver Met-tRNA$_i^{Met}$ to the 40S ribosome to initiate translation	[31]
hOLA1	*Homo sapiens*	Heat shock protein 70 (HSP70)	OLA1 can interfere with the binding and function of the E3 ligase CHIP to HSP70, leading to the stabilization of HSP70, and response to heat shock	[32]
hOLA1	*Homo sapiens*	Superoxide dismutase 2 (SOD2)	OLA1 deficiency can enhance CHIP affinity for HSP70-SOD2 complexes, facilitating SOD2 degradation, supporting OLA1 plays a role in response to mitochondrial oxidative stress	[33]

hOLA1 (human Obg-like ATPase1) (~45 kDa) is a human ortholog of YchF that is expressed in the cytoplasm [30,34]. The *hOLA1* overexpression cells showed increased sensitivity to oxidant-induced cytotoxicity. Conversely, *hOLA1*-knockdown cells conferred tolerance to oxidizing agents, such as tert-butyl hydroperoxide (tBH) and diamide, and *hOLA1*-knockdown cells demonstrated reduced cellular reactive oxygen species (ROS) production [34].

Escherichia coli YchF expression is growth phase-dependent and down-regulated under oxidative stress conditions [30]. *E. coli YchF* overexpression enhanced cellular sensitivity to H_2O_2-induced oxidative stress, while the *E. coli YchF* deletion strain displayed increased resistance against H_2O_2 and diamide [18]. Although *E. coli* YchF physically interacts with the *E. coli* catalase KatG and *E. coli YchF* overexpression inhibits KatG enzyme activity in vivo, there is no effect on KatG enzyme activity in the presence of the purified *E. coli* YchF in vitro, suggesting that the reduced catalase activity should be an indirect effect in vivo (Figure 3; Table 2) [18].

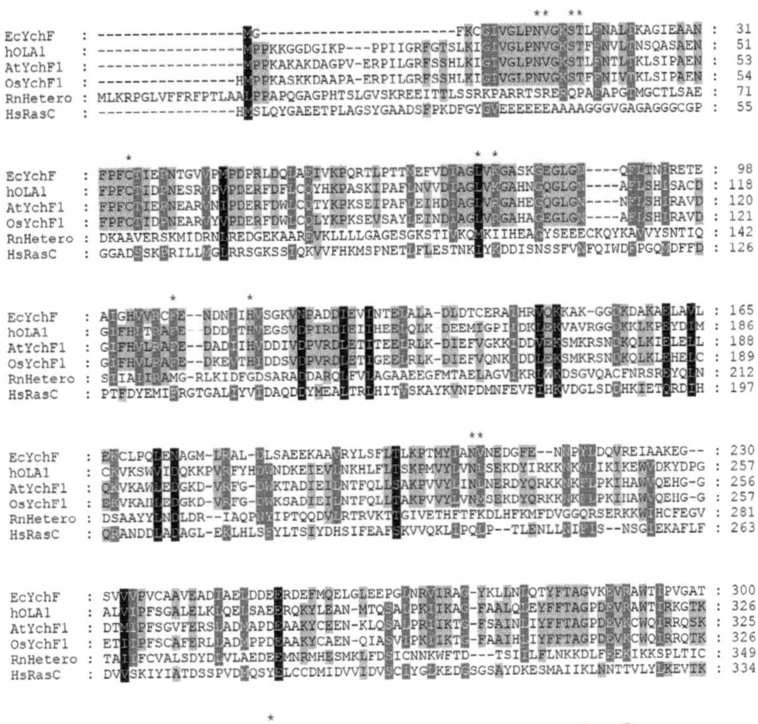

Figure 3. Sequence alignments of *E. coli* YchF (EcYchF), hOLA1, OsYchF1, AtYchF1, *H. sapiens* Ras-related G protein C and *R. norvegicus* heterotrimeric G protein α-subunit using software Jalview version 1.6 (https://www.jalview.org). *E. coli* YchF NCBI Protein code is VWQ02248.1, hOLA1 NCBI Protein code is NP_037473.3, OsYchF1 NCBI Protein code is BAD03576.1, AtYchF1 NCBI Protein code is Q9SA73.1, human Ras-related G protein C (HsRas C) NCBI Protein code is NP_071440.1, and *R. norvegicus* heterotrimeric G protein α-subunit (RnHetero) NCBI Protein code is XP_010846404.1. The conserved amino acid residues are marked in the dark, and the more conserved amino acid residues are much darker, and the specific amino acid residues in Table 1 were marked with *.

Moreover, *E. coli* YchF functions as a redox-regulated monomer-dimer equilibrium through a conserved cysteine residue 35 within the *E. coli* YchF nucleotide-binding site (Table 1) [16]. The *E. coli* YchF dimer shows a low ATPase activity, but the *E. coli* YchF monomer displays significantly increased *E. coli* YchF ATPase activity (Figure 3; Table 2) [35]. Thioredoxin 1 (TrxA) maintains the redox balance in vivo and directly interacts with the G domain and coiled-coil domain of *E. coli* YchF to dissociate *E. coli* YchF dimer. Interestingly, wild-type *E. coli* cells effectively outcompete the *E. coli* YchF deletion strain, indicating that *E. coli* YchF might influence *E. coli* cell growth, but the mechanism is unclear [30].

3.2. *YchF Is Crucial for Environmental Stress Response*

A fatal marine bacterium *Vibrio vulnificus,* ortholog of YchF, elicits macrophage cytotoxicity. It shows a significant negative effect of macrophage cytotoxicity on iron-overloaded mice through the *rtxA1* pathway that stimulates cytotoxicity to macrophages [31,36]. The *V. vulnificus YchF* deletion strain displayed retarded growth and reduced transcription level of the *rtxA1* gene [31,36]. In addition, *Propionibacterium acidipropionici* ortholog of YchF is crucial for the regulation of propionic acid tolerance [31,36].

The expression of *hOLA1*, namely *DNA damage-regulated overexpressed in cancer 45* (*DOC45*), was strongly down-regulate by DNAdamage-inducing agents, such as etoposide, doxorubicin (adriamycin), and ionizing and UV radiation, but not endoplasmic reticulumstress-inducing agents [20]. Compared with normal human cells, *hOLA1* expression is notably upregulated in established colon cancer cells at both mRNA and protein levels [20]. *hOLA1*-knockdown human colon cancer cells show a negative impact on cell proliferation and hypersensitivity to Adriamycin-induced cell death [20,34,37].

3.3. *YchF Bridges Protein Biosynthesis and Degradation*

YchF anchors to ribosomes and polysomes, suggesting that YchF is involved in protein biosynthesis (Table 2) [17,24]. The ribosome is likely in contact with the N-terminal G domain of YchF [34]. Consistently, the 70 S ribosomal subunit is able to enhance *E. coli* YchF ATPase activity, although *E. coli* YchF hardly influences the assembly and steady-state amounts of ribosomes [38]. *E. coli* YchF preferentially binds to the translation initiation factor 3 (IF3) and several ribosomal proteins at the surface of the 30 S ribosomal particle, while the interaction of *E. coli* YchF with 50 S ribosomal particle seems probably only transient [17,24].

The percentage of leaderless mRNAs is only 0.7% in *E. coli* BW25113 under normal growth conditions, but the ribonuclease MazF generates leaderless mRNAs by cleaving off the Shine-Dalgarno (SD) sequence close to the start-codon upon environmental stress [39,40]. The MazF-generated leaderless mRNA modulation is necessary for bacterial survival under environmental stress [41]. Compared with wild-type *E. coli* cells, the *E. coli YchF* deletion strain showed increased resistance against MazF-generated leaderless mRNAs [30]. In other words, *E. coli* YchF suppresses the translation of MazF-processed mRNAs upon stress conditions and declines the resistance towards the endoribonuclease [30] (Figure 3; Table 2). In addition, the *E. coli YchF* deletion strain demonstrated increased resistance to hydroxyurea (HU), a ribonucleotide reductase inhibitor, and fusidic acid, an elongation factor G (EF-G) inhibitor [42,43].

A tandem-affinity purification and mass spectrometry (TAP-MS) approach shows the interaction between yeast *Saccharomyces cerevisiae* ortholog of YchF (YBR025C) and eukaryotic translation elongation factor 1 (eEF1), committed with protein translation (Table 2) [28,29]. Additionally, hOLA1 interacts with eukaryotic initiation factor 2 (eIF2) mediates ribosomal recruitment of the initiator methionyl-tRNA (tRNAi) and interferes with the eIF2-mediated formation of a ternary complex with GTP and tRNAi [26] (Figure 3; Table 2).

In the protozoan *Trypanosoma cruzi*, the ortholog of YchF (~44.3 kDa) is associated with not only ribosomal particles and polysomes but also proteasomal subunits allowing protein degradation by the ubiquitin-proteasome pathway (Table 2) [26]. Immunoprecipitation assays exhibited that *T. cruzi* YchF co-sediments with the non-ATPase subunit RPN10 of the *T. cruzi* proteasome, which might mediate damaged protein degradation during protein biosynthesis under stress conditions [26]. An integrated mass spectrometry-based proteomic approach also indicated that the 26S proteasome links with *S. cerevisiae* YchF (YBR025C) in yeast [27]. Due to the structural similarity of the TGS domain to ubiquitin-like proteins, the C-terminal TGS domain of YchF is a potential candidate for the interaction between YchF and the subunits of the proteasome [44]. Noticeably, the absence of *T. cruzi YchF* restrains the cellular growth of *T. brucei* as well as the procyclic forms of the parasite [26]. As hOLA1 and *E. coli* YchF, *T. cruzi* YchF also bind and hydrolyzes ATP more efficiently than GTP [14,17].

3.4. YchF Is a Key Molecule in Maintaining Proteostasis

The well-known heat-shock response is a major strategy towards environmental stimuli by the rapid biosynthesis of the molecular chaperone heat-shock proteins [32]. Heat-shock proteins are essential for maintaining intracellular homeostasis by assisting in the damaged proteins [32]. Heat-Shock Protein 70 (HSP70) plays a key role in multiple primary human cancers, and high expression of HSP70 is related to poor tumor progression [33,37]. hOLA1 interacts with the C-terminal variable domain of HSP70 to prevent contact with the C-terminus of Hsp70-binding protein (CHIP), an E3 ubiquitin ligase for HSP70, thereby inhibiting HSP70 from the CHIP-mediated ubiquitination. Thus, hOLA1 stabilizes HSP70 to improve survival under stress conditions [32]. Additionally, Hsp70 is also a molecular chaperone for mitochondrial superoxide dismutase 2 (SOD2), which is responsible for keeping normal mitochondrial reactive oxygen species (ROS) [33]. hOLA1 directly recruits Hsp70 and SOD2 to hinder them from ubiquitin-dependent protein degradation under stress conditions [33]. In conclusion, YchF controls multiple proteostatic mechanisms in response to environmental stresses.

4. OsYchF1/AtYchF1 and Its Activator OsGAP1/AtGAP1 in Plants

In nature, plants are often exposed to various environmental stresses during growth and development, including flooding, drought, salt, cold, insect herbivores, and microbes [45,46]. Plants have to evolve sophisticated mechanisms to guard themselves against these environmental challenges [45,47]. Rice (*O. sativa*) GTPase-activating protein 1 (OsGAP1), a C2 domain-containing protein involved in plant defense response pathway, was originally identified using suppression subtraction hybridization (SSH) of a Xa14 rice cDNA library derived from a rice line harboring the *Xa14* resistance gene against the bacterial pathogen *Xanthomonas oryzae* pv. *oryzae* (*Xoo*) [1]. OsGAP1 was constructed as a bait to capture the prey OsYchF1 by yeast two-hybrid assay. The transgenic *Arabidopsis thaliana* ectopically overexpressing *OsGAP1* showed increased resistance with upregulating expressions of both salicylic acid (SA)-related (*PR1* and *PR2*) and jasmonic acid (JA)-related (*Thi2.1* and *PDF1.2*) defense marker genes on *Pst* DC3000 was dependent on [1]. Furthermore, the resistant effects of *OsGAP1* on *Pst* DC3000 are dependent on the key plant biotic stress response regulator *NONEXPRESSOR OF PATHOGENESIS-RELATED GENES 1* (*NPR1*) [1]. *OsGAP1* ectopically overexpressed in the *A. thaliana npr1-3* mutant never showed increased resistance towards *Pst* DC3000 [1].

OsYchF1 almost utilizes ATP and GTP equally, unlike protozoan, bacterium, and human YchF orthologs that give priority to ATP over GTP [1]. OsGAP1 significantly enhances OsYchF1 ATPase and GTPase activities and turns OsYchF1 into the inactive GDP or ADP-bound state [1]. Moreover, OsGAP1 might control the subcellular localization of OsYchF1 by recruiting cytosolic OsYchF1 to the intracellular plasma membrane subjected to wounding treatment [1]. OsYchF1 and its activating protein OsGAP1 play opposite roles in response to environmental stimuli (Figure 3; Table 2). On the one hand, the *OsYchF1* overexpressors are sensitive to the bacterial pathogen *Pseudomonas syringae* pv. *tomato* DC3000 (*Pst* DC3000), but the *A. thaliana YchF* (*AtYchF1*) knockout mutant and OsGAP1 overexpressors confer tolerance to the bacterial pathogen in *A. thaliana* [1,48]. On the other hand, the overexpression of *OsYchF1* and *AtYchF1* in transgenic *A. thaliana* results in decreased resistance to high salinity-induced oxidative stress, while the overexpression of *OsGAP1* or *AtGAP1* (*OsGAP1* ortholog in *A. thaliana*) and the *AtYchF1* knockout mutant alleviate salt stress [21].

In order to dissect the interaction of OsYchF1 with OsGAP1, firstly, site-directed mutagenesis identifies three clusters (D23, D28; R117, N119, E123, E124; R141, R143, E146, E149) of OsGAP1 surface amino acid residues that are essential for binding to phospholipids, which play an important role in enhancing defense responses. Additionally, the effects of OsGAP1 on high salinity tolerance are dependent on the interaction between the other two clusters (L5, L8, T58, S60, and Ser-60; K37, K39, K41, R43) of OsGAP1 and OsYchF1 [19,20]. Secondly, a recent study explains that four critical amino acid residues (Lys-325, His-334, Glu-345, and Glu-354) in the OsYchF1 TGS domain are required for the interaction of OsGAP1 with OsYchF1 [19].

Slot blot analysis demonstrates that the OsYchF1 TGS domain interacts with the 26S RNA in rice, suggesting that OsYchF1 is committed to protein biosynthesis as well [1]. Additionally, recent co-crystallization and biochemical data showed that AtYchF1 in the complex with ppGpp inhibits the interaction of AtYchF1 with other molecules, including ATP, GTP, and 26S rRNA [23]. The available data indicate that ppGpp works as an alarmone in response to environmental stimuli, and the concentration of ppGpp in the cytoplasm can increase to the millimolar level upon stress conditions [3,23,49]. Most importantly, the accumulation of ppGpp attenuates plant growth and development [23]. In conclusion, *AtYchF1* might be a critical regulatory factor in controlling the cytosolic ppGpp-mediated growth inhibition in plants (Table 2).

5. Conclusions and Outlook

YchF subfamily universally exists in both bacteria and eukarya except archaea [1,8]. The N-terminal G domain consists of five motifs that are highly conserved among all P-loop GTPases [14,50], of which G4 and G5 motifs determine specific ATPase or GTPase activities. YchF plays a critical role in regulating growth and stress responses among different organisms and life processes. In the current review, we have listed the important and conserved amino acid residues of YchF in not only Table 1 but also the interactive partner proteins in Table 2.

The unique YchF is probably a guanosine or adenosine nucleotide-dependent translational factor associated with the ribosomal particles and subunits of the proteasome, potentially bridging the protein biosynthesis [1,24,51] and ubiquitin-dependent protein degradation to maintain proteostasis [26,32,33]. YchF is also involved in life response to environmental challenges by recruiting many partner proteins [1] (Figure 4). In prokaryotes, YchF might function as a GTP-dependent translation factor, participating in the translation process as part of the nucleoprotein complex [17,24,30,38] and being involved in oxidative stress response [18]. In yeast, as a representative of eukaryotes, a YchF homologous YBR025c is induced by H_2O_2 and participates in the degradation of damaged proteins by interacting with the 26S proteasome in response to oxygen stress [27]. The human Obg-like ATPase1 (hOLA1) is a human homolog of YchF that is overexpressed in several human malignancies and acts as a negative regulator of multiple oxidants [34,37].

OsYchF1 is a novel unconventional G protein in rice. The molecular mechanisms of *OsYchF1* in rice still remain largely unknown. The gain-of-function *OsYchF1* overexpression transgenic lines and loss-of-function *OsYchF1* knockout or knockdown mutants have not been constructed and monitored in rice, and the agronomic traits of *OsYchF1* need to be observed. Moreover, YchF is not only related to stress responses but also appears to influence the metabolic processes in diverse species, which should be delineated and clarified in the future.

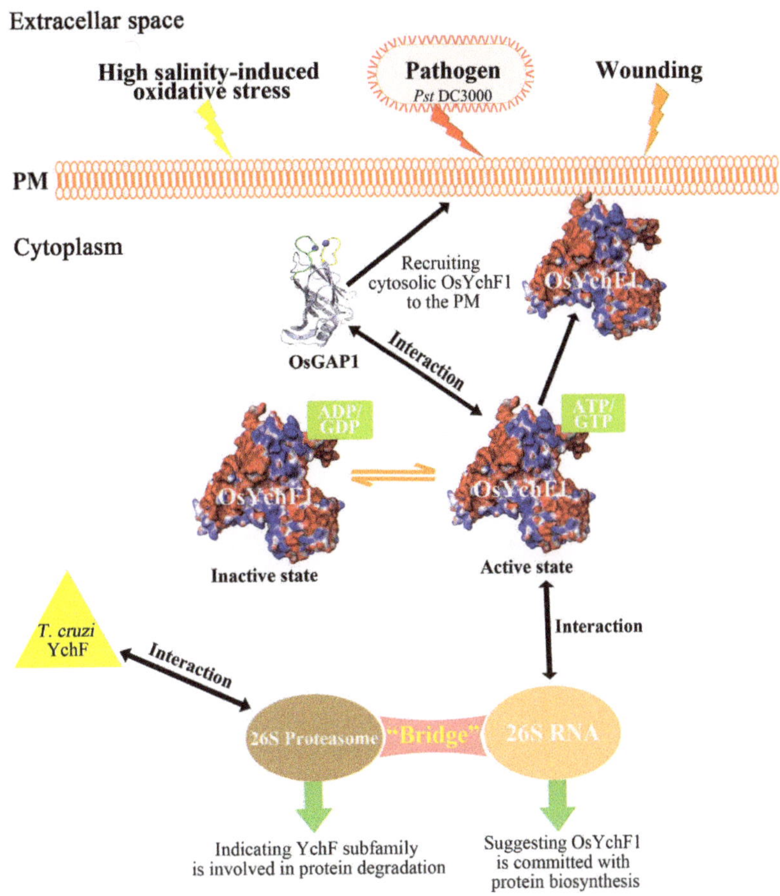

Figure 4. Working model to explain the structure-function relationship of YchF. The interaction of OsYchF1 (PDB code: 5EE0) and its activating protein OsGAP1 (PDB code: 4RJ9) participate in stress response. OsYchF1 is activated when binding ATP/GTP and inactivated when binding ADP/GDP. The possible role of members of the Ychf subfamily might be involved in protein balance. OsYchF1 can interact with OsGAP1 and 26S RNA, while *T. cruzi* Ychf can interact with 26S proteasome, suggesting that the Ychf subfamily may act as a bridge between protein synthesis and degradation.

Author Contributions: Conceptualization, Z.L., R.L. and R.M.; writing—review and editing, Z.L., R.L., Z.H., Y.L., L.G., S.H., Y.M. and R.M.; funding acquisition, R.M. All authors have read and agreed to the published version of the manuscript.

Funding: We are grateful for grant support from the Department of Science and Technology of Fujian Province (KTP22112A-2021; KTP22288A-2022; 2023J01133697), and Research Grant of FAFU (114-722022003).

Institutional Review Board Statement: Not applicable.

Informed Consent Statement: Not applicable.

Data Availability Statement: Not applicable.

Acknowledgments: We thank Rui Miao's PhD thesis in the Chinese University of Hong Kong.

Conflicts of Interest: The authors declare no conflict of interest.

References

1. Cheung, M.Y.; Xue, Y.; Zhou, L.; Li, M.W.; Sun, S.S.M.; Lam, H.M. An ancient P-loop GTPase in rice is regulated by a higher plant-specific regulatory protein. *J. Biol. Chem.* **2010**, *285*, 37359–37369. [CrossRef] [PubMed]
2. Yung, Y.L.; Cheung, M.Y.; Miao, R.; Fong, Y.H.; Li, K.P.; Yu, M.H.; Chye, M.L.; Wong, K.B.; Lam, H.M. Site-directed mutagenesis shows the significance of interactions with phospholipids and the G-protein OsYchF1 on the physiological functions of the rice GTPase-Activating Protein 1 (OsGAP1). *J. Biol. Chem.* **2015**, *290*, 23984–23996. [CrossRef] [PubMed]
3. Buglino, J.; Shen, V.; Hakimian, P.; Lima, C.D. Structural and biochemical analysis of the Obg GTP binding protein. *Structure* **2002**, *10*, 1581–1592. [CrossRef] [PubMed]
4. Assmann, S.M. Heterotrimeric and unconventional GTP binding proteins in plant cell signaling. *Plant Cell* **2002**, *14* (Suppl. S1), S355–S373. [PubMed]
5. Wennerberg, K.; Rossman, K.L.; Der, C.J. The Ras superfamily at a glance. *J. Cell Sci.* **2005**, *118 Pt 5*, 843–846.
6. Milligan, G.; Kostenis, E. Heterotrimeric G-proteins: A short history. *Br. J. Pharmacol.* **2009**, *147*, 46–55. [CrossRef]
7. Sprang, S.R. G protein mechanisms: Insights from structural analysis. *Annu. Rev. Biochem.* **1997**, *66*, 639–678. [CrossRef]
8. Leipe, D.D.; Wolf, Y.I.; Koonin, E.V.; Aravind, L. Classification and evolution of P-loop GTPases and related ATPases. *J. Mol. Biol.* **2002**, *317*, 41–72.
9. Luo, M.; Han, Z.; Huang, G.; Li, R.; Liu, Y.; Lu, J.; Liu, L.; Miao, R. Structural comparison of unconventional G protein YchF with heterotrimeric G protein and small G protein. *Plant Signal. Behav.* **2022**, *17*, 2024405. [CrossRef]
10. Oldham, W.M.; Hamm, H.E. Heterotrimeric G protein activation by G-protein-coupled receptors. *Nat. Rev. Mol. Cell. Biol.* **2008**, *9*, 60–71. [CrossRef]
11. Ma, Y.; Dai, X.; Xu, Y.; Luo, W.; Zheng, X.; Zeng, D.; Pan, Y.; Lin, X.; Liu, H.; Zhang, D.; et al. COLD1 confers chilling tolerance in rice. *Cell* **2015**, *160*, 1209–1221. [CrossRef] [PubMed]
12. Maruta, N.; Trusov, Y.; Jones, A.M.; Botella, J.R. Heterotrimeric G Proteins in Plants: Canonical and Atypical Gα Subunits. *Int. J. Mol. Sci.* **2021**, *22*, 11841. [CrossRef] [PubMed]
13. Reiner, D.J.; Lundquist, E.A. Small GTPases. *WormBook* **2018**, *16*, 1–65. [CrossRef] [PubMed]
14. Koller-Eichhorn, R.; Marquardt, T.; Gail, R.; Wittinghofer, A.; Kostrewa, D.; Kutay, U.; Kambach, C. Human OLA1 defines an ATPase subfamily in the Obg family of GTP-binding proteins. *J. Biol. Chem.* **2007**, *282*, 19928–19937. [CrossRef] [PubMed]
15. Rosler, K.S.; Mercier, E.; Andrews, I.C.; Wieden, H.J. Histidine 114 Is Critical for ATP Hydrolysis by the Universally Conserved ATPase YchF. *J. Biol. Chem.* **2015**, *290*, 18650–18661. [CrossRef] [PubMed]
16. Hannemann, L.; Suppanz, I.; Ba, Q.; MacInnes, K.; Drepper, F.; Warscheid, B.; Koch, H.G. Redox Activation of the Universally Conserved ATPase YchF by Thioredoxin 1. *Antioxid. Redox Signal.* **2016**, *24*, 141–156. [CrossRef]
17. Tomar, S.K.; Kumar, P.; Prakash, B. Deciphering the catalytic machinery in a universally conserved ribosome binding ATPase YchF. *Biochem. Biophys. Res. Commun.* **2011**, *408*, 459–464. [CrossRef]
18. Wenk, M.; Ba, Q.; Erichsen, V.; MacInnes, K.; Wiese, H.; Warscheid, B.; Koch, H.G. A universally conserved ATPase regulates the oxidative stress response in Escherichia coli. *J. Biol. Chem.* **2012**, *287*, 43585–43598. [CrossRef]
19. Cheung, M.Y.; Ngo, J.C.; Chen, Z.; Jia, Q.; Li, T.; Gou, Y.; Wang, Y.; Lam, H.M. A structure model explaining the binding between a ubiquitous unconventional G-protein (OsYchF1) and a plant-specific C2-domain protein (OsGAP1) from rice. *Biochem. J.* **2020**, *477*, 3935–3949. [CrossRef]
20. Sun, H.; Luo, X.; Montalbano, J.; Jin, W.; Shi, J.; Sheikh, M.S.; Huang, Y. DOC45, a novel DNA damage-regulated nucleocytoplasmic ATPase that is overexpressed in multiple human malignancies. *Mol. Cancer Res.* **2010**, *8*, 57–66. [CrossRef]
21. Cheung, M.Y.; Li, M.W.; Yung, Y.L.; Wen, C.Q.; Lam, H.M. The unconventional P-loop NTPase OsYchF1 and its regulator OsGAP1 play opposite roles in salinity stress tolerance. *Plant Cell Environ.* **2013**, *36*, 2008–2020. [CrossRef] [PubMed]
22. Cheung, M.Y.; Li, X.; Miao, R.; Fong, Y.H.; Li, K.P.; Yung, Y.L.; Yu, M.H.; Wong, K.B.; Chen, Z.; Lam, H.M. ATP binding by the P-loop NTPase OsYchF1 (an unconventional G protein) contributes to biotic but not abiotic stress responses. *Proc. Natl. Acad. Sci. USA* **2016**, *113*, 2648–2653. [CrossRef] [PubMed]
23. Cheung, M.Y.; Li, X.; Ku, Y.S.; Chen, Z.; Lam, H.M. Co-crystalization reveals the interaction between AtYchF1 and ppGpp. *Front. Mol. Biosci.* **2022**, *9*, 1061350. [CrossRef] [PubMed]
24. Becker, M.; Gzyl, K.E.; Altamirano, A.M.; Vuong, A.; Urban, K.; Wieden, H.J. The 70S ribosome modulates the ATPase activity of *Escherichia coli* YchF. *RNA Biol.* **2012**, *9*, 1288–1301. [CrossRef] [PubMed]
25. Brandon, H.E. *Biophysical Studies of the Universally Conserved NTPases HflX and YchF*; University of Lethbridge: Lethbridge, AB, Canada, 2021.
26. Gradia, D.F.; Rau, K.; Umaki, A.C.; de Souza, F.S.; Probst, C.M.; Correa, A.; Holetz, F.B.; Avila, A.R.; Krieger, M.A.; Goldenberg, S.; et al. Characterization of a novel Obg-like ATPase in the protozoan *Trypanosoma cruzi*. *Int. J. Parasitol.* **2009**, *39*, 49–58. [CrossRef]
27. Guerrero, C.; Tagwerker, C.; Kaiser, P.; Huang, L. An integrated mass spectrometry-based proteomic approach: Quantitative analysis of tandem affinity-purified in vivo cross-linked protein complexes (QTAX) to decipher the 26 S proteasome-interacting network. *Mol. Cell Proteom.* **2005**, *5*, 366–378. [CrossRef] [PubMed]
28. Teplyakov, A.; Obmolova, G.; Chu, S.Y.; Toedt, J.; Eisenstein, E.; Howard, A.J.; Gilliland, G.L. Crystal structure of the YchF protein reveals binding sites for GTP and nucleic acid. *J. Bacteriol.* **2003**, *185*, 4031–4037. [CrossRef]
29. Gavin, A.C.; Bösche, M.; Krause, R.; Grandi, P.; Marzioch, M.; Bauer, A.; Schultz, J.; Rick, J.M.; Michon, A.M.; Cruciat, C.M.; et al. Functional organization of the yeast proteome by systematic analysis of protein complexes. *Nature* **2002**, *415*, 141–147. [CrossRef]

30. Landwehr, V.; Milanov, M.; Angebauer, L.; Hong, J.; Jüngert, G.; Hiersemenzel, A.; Siebler, A.; Schmit, F.; Öztürk, Y.; Dannenmaier, S.; et al. The Universally Conserved ATPase YchF Regulates Translation of Leaderless mRNA in Response to Stress Conditions. *Front. Mol. Biosci.* **2021**, *8*, 643696. [CrossRef]
31. Chen, H.; Song, R.; Wang, G.; Ding, Z.; Yang, C.; Zhang, J.; Zeng, Z.; Rubio, V.; Wang, L.; Zu, N.; et al. OLA1 regulates protein synthesis and integrated stress response by inhibiting eIF2 ternary complex formation. *Sci. Rep.* **2015**, *5*, 13241. [CrossRef]
32. Mao, R.F.; Rubio, V.; Chen, H.; Bai, L.; Mansour, O.C.; Shi, Z.Z. OLA1 protects cells in heat shock by stabilizing HSP70. *Cell Death Dis.* **2013**, *4*, e491. [CrossRef]
33. Schultz, A.; Olorundami, O.A.; Teng, R.J.; Jarzembowski, J.; Shi, Z.Z.; Kumar, S.N.; Pritchard KJr Konduri, G.G.; Afolayan, A.J. Decreased OLA1 (Obg-Like ATPase-1) Expression Drives Ubiquitin-Proteasome Pathways to Downregulate Mitochondrial SOD2 (Superoxide Dismutase) in Persistent Pulmonary Hypertension of the Newborn. *Hypertension* **2019**, *74*, 957–966. [CrossRef]
34. Zhang, J.; Rubio, V.; Lieberman, M.W.; Shi, Z.Z. OLA1, an Obg-like ATPase, suppresses antioxidant response via nontranscriptional mechanisms. *Proc. Natl. Acad. Sci. USA* **2009**, *106*, 15356–15361. [CrossRef]
35. Guan, N.; Shin, H.D.; Chen, R.R.; Li, J.; Liu, L.; Du, G.; Chen, J. Understanding of how Propionibacterium acidipropionici respond to propionic acid stress at the level of proteomics. *Sci. Rep.* **2014**, *4*, 6951. [CrossRef]
36. Chen, Y.C.; Chung, Y.T. A conserved GTPase YchF of Vibrio vulnificus is involved in macrophage cytotoxicity, iron acquisition, and mouse virulence. *Int. J. Med. Microbiol.* **2011**, *301*, 469–474. [CrossRef]
37. Dong, Y.; Yin, A.; Xu, C.; Jiang, H.; Wang, Q.; Wu, W.; Guo, S. OLA1 is a potential prognostic molecular biomarker for endometrial cancer and promotes tumor progression. *Oncol. Lett.* **2021**, *22*, 576. [CrossRef]
38. Gibbs, M.R.; Moon, K.M.; Warner, B.R.; Chen, M.; Bundschuh, R.; Foster, L.J.; Fredrick, K. Functional Analysis of BipA in *E. coli* Reveals the Natural Plasticity of 50S Subunit Assembly. *J. Mol. Biol.* **2020**, *432*, 5259–5272. [CrossRef]
39. Beck, H.J.; Moll, I. Leaderless mRNAs in the Spotlight: Ancient but Not Outdated! *Microbiol. Spectr.* **2018**, *6*, RWR-0016-2017. [CrossRef]
40. Vesper, O.; Amitai, S.; Belitsky, M.; Byrgazov, K.; Kaberdina, A.C.; Engelberg-Kulka, H.; Moll, I. Selective translation of leaderless mRNAs by specialized ribosomes generated by MazF in *Escherichia coli*. *Cell* **2011**, *147*, 147–157. [CrossRef]
41. Amitai, S.; Kolodkin-Gal, I.; Hananya-Meltabashi, M.; Sacher, A.; Engelberg-Kulka, H. *Escherichia coli* MazF leads to the simultaneous selective synthesis of both "death proteins" and "survival proteins". *PLoS Genet.* **2009**, *5*, e1000390. [CrossRef]
42. Rodnina, M.V.; Wintermeyer, W. Protein elongation, Co-translational folding and targeting. *J. Mol. Biol.* **2016**, *428*, 2165–2185. [CrossRef]
43. Spivak, J.L.; Hasselbalch, H. Hydroxycarbamide: A user's guide for chronic myeloproliferative disorders. *Expert Rev. Anti Cancer Ther.* **2011**, *11*, 403–414. [CrossRef]
44. Iyer, L.M.; Burroughs, A.M.; Aravind, L. The prokaryotic antecedents of the ubiquitin-signaling system and the early evolution of ubiquitin-like beta-grasp domains. *Genome Biol.* **2006**, *7*, R60. [CrossRef]
45. Miao, R.; Russinova, E.; Rodriguez, P.L. Tripartite hormonal regulation of plasma membrane H+-ATPase activity. *Trends Plant Sci.* **2022**, *27*, 588–600. [CrossRef]
46. Miao, R.; Yuan, W.; Wang, Y.; Garcia-Maquilon, I.; Dang, X.; Li, Y.; Zhang, J.; Zhu, Y.; Rodriguez, P.L.; Xu, W. Low ABA concentration promotes root growth and hydrotropism through relief of ABA INSENSITIVE 1-mediated inhibition of plasma membrane H+-ATPase 2. *Sci. Adv.* **2021**, *7*, eabd4113. [CrossRef]
47. Miao, R.; Wang, M.; Yuan, W.; Ren, Y.; Li, Y.; Zhang, N.; Zhang, J.; Kronzucker, H.J.; Xu, W. Comparative analysis of Arabidopsis ecotypes reveals a role for brassinosteroids in root hydrotropism. *Plant Physiol.* **2018**, *176*, 2720–2736. [CrossRef]
48. Cheung, M.Y.; Zeng, N.Y.; Tong, S.W.; Li, W.Y.; Xue, Y.; Zhao, K.J.; Wang, C.; Zhang, Q.; Fu, Y.; Sun, Z.; et al. Constitutive expression of a rice GTPase-activating protein induces defense responses. *New Phytol.* **2008**, *179*, 530–545. [CrossRef]
49. Ito, D.; Kawamura, H.; Oikawa, A.; Ihara, Y.; Shibata, T.; Nakamura, N.; Asano, T.; Kawabata, S.I.; Suzuki, T.; Masuda, S. ppGpp functions as an alarmone in metazoa. *Commun. Biol.* **2020**, *3*, 671. [CrossRef]
50. Caldon, C.E.; Yoong, P.; March, P.E. Evolution of a molecular switch: Universal bacterial GTPases regulate ribosome function. *Mol. Microbiol.* **2001**, *41*, 289–297. [CrossRef]
51. Cruz-Vera, L.R.; Galindo, J.M.; Guarneros, G. Transcriptional analysis of the gene encoding peptidyl-tRNA hydrolase in Escherichia coli. *Microbiology* **2002**, *148 Pt 11*, 3457–3466. [CrossRef]

life

Article

The Effect of Hormone Types, Concentrations, and Treatment Times on the Rooting Traits of *Morus* 'Yueshenda 10' Softwood Cuttings

Jiajia Sun [1], Hongyun Li [2], Hanlei Chen [1], Tiantian Wang [1], Jin'e Quan [1,*] and Huitao Bi [1,*]

1 College of Forest, Henan Agricultural University, Zhengzhou 450002, China; dearperi@163.com (J.S.); 18336253160@163.com (H.C.); wangtiantian04@163.com (T.W.)
2 Management Office, Beijing Tiantan Park, Beijing 100061, China; hyunli926@126.com
* Correspondence: quanjine@163.com (J.Q.); bihuitao@126.com (H.B.); Tel.: +86-177-4462-4393 (J.Q.); +86-135-9880-7967 (H.B.)

Abstract: Enhancing the capacity of fruit trees to propagate via cuttings is an important endeavor for the high-quality development of the fruit industry. Optimizing the conditions for the cutting propagation of mulberry seedlings is an important factor that influences the industrial production of this plant; however, the currently used mulberry breeding technology system is not mature. In this experiment, an orthogonal design was used to intercept semi-woody shoots of Yueshenda 10 as cuttings and set different hormone concentrations (200, 500, 800, and 1000 mg/L), different hormone types (NAA, IBA, IAA, and ABT-1), and different soaking times (10, 30, 60, and 120 min) for cuttings. The effects of the three factors on the rooting of mulberry cuttings were investigated by soaking the cuttings in clean water for 10 min as a control. The results showed that the primary and secondary order of the three factors affecting the rooting rate of cuttings was hormone concentration > hormone type > soaking time, and the concentration of exogenous hormones had a significant impact on all rooting indicators ($p < 0.05$). In addition, the rooting rate (66.24%), average number of roots (7.54 roots/plant), and rooting effect index (4.23) of Yueshenda 10 cuttings reached the optimal level when soaked with 800 mg/L ABT-1 for 30 min. The longest root length (10.20 cm) and average root length (4.44 cm) of cuttings achieved the best results when soaked with 800 mg/L NAA for 60 min and 500 mg/L NAA for 30 min, respectively. On balance, it is considered that the preferred solution is to soak the cuttings of Yueshenda 10 with 800 mg/L ABT1 solution for 0.5 h.

Keywords: mulberry; softwood cutting; growth regulator; orthogonal experiment; root morphological index

Citation: Sun, J.; Li, H.; Chen, H.; Wang, T.; Quan, J.; Bi, H. The Effect of Hormone Types, Concentrations, and Treatment Times on the Rooting Traits of *Morus* 'Yueshenda 10' Softwood Cuttings. *Life* **2023**, *13*, 1032. https://doi.org/10.3390/life13041032

Academic Editor: Balazs Barna

Received: 20 March 2023
Revised: 14 April 2023
Accepted: 14 April 2023
Published: 17 April 2023

1. Introduction

Mulberry (*Morus*) belongs to the family Moraceae and is a perennial woody plant. The fruit mulberry is a type of mulberry tree chosen from the Moraceae family of plants specifically for producing appropriate mulberries for either direct consumption or processing [1]. The production, life, and ecological functions of mulberry fruits, leaves, and branches make fruit mulberry not only an economically valuable forest tree but also a tree species with significant value for environmental management and ecological optimization [2].

The need for healthy, green, and efficient development of the fruit mulberry industry calls for improvement of the breeding and cultivation ability, starting with the provision of sufficient high-quality seedlings. Currently, there are three commonly used asexual reproduction techniques for mulberry seedling cultivation: tissue culture, grafting, and cutting [3]. When compared to tissue culture, which has high prerequisites, and grafting, which has a lengthy cycle [4], cutting propagation, particularly softwood cutting, can propagate substantial quantities while retaining the superior characteristics of the parent plant and preventing deterioration. Furthermore, it is straightforward to manage and promote at a lower cost [5–7]. Therefore, cutting propagation has become one of the main seedling breeding methods that is ideal for the large-scale planting needs of mulberry orchards [8–10].

Investigations into the effects of plant growth regulators on the rooting of cuttings dates back to 1934 when Went explored the types, concentrations, and treatment times of exogenous hormones [11,12]. The formation of adventitious roots is a process regulated by various factors, including the lignification degree of the cutting, the season of cutting, the rooting environment, and exogenous hormones [13]. Among these factors, the cutting medium and hormone application are two essential external factors affecting the rooting and root quality of cuttings [14].

In the 1930s, indole-3-acetic acid (IAA) was proven to be effective in promoting the formation of adventitious root [15]. Indole-3-butyric acid (IBA) has become the most commonly used natural auxin for plant reproduction in horticulture and forestry because of its stability and effectiveness in promoting rooting of stem cuttings [16,17]. A-naphthylacetic acid (NAA) plays an important role in promoting cell division and expansion, inducing adventitious root formation, increasing fruit setting, and preventing fruit drop [18]. Different phytohormone concentrations also result in different plant growth, development, and responses to external stimuli [19,20].

Currently, these plant growth regulators are widely used in cutting seedling cultivation. ABT1 and ABT2 rooting powders have been reported to increase rooting rates up to 95% in green stem cuttings [21]. For *Morus alba* L., the application of 0.10% to 0.13% IBA and ABT rooting powder No. 6 resulted in a survival rate of over 81% [22–24]. Other studies have demonstrated that the combination of NAA and IAA may be better compared with NAA alone in the propagation of mulberry [25,26]. Deng et al. compared the effects of IAA-3-acetic acid and ABT1 rooting powder alone and in combination and found that the combination of the two had more benefits compare to the use of either of the two alone [27].

These studies have demonstrated that auxins can promote the growth of adventitious rooting of cuttings. The aim of this study was to determine the most suitable hormone combination for softwood cutting propagation of mulberry to provide a basis for developing high-quality mulberry seedlings within a short time to meet the high market demand. The results are expected to provide a theoretical basis and technical guidance for the propagation of mulberry through softwood cutting. Therefore, a green stem cutting experiment was conducted on Yueshenda 10, and the effects of different influencing factors on the rooting characteristics of Yueshenda 10 softwood cuttings were compared using an orthogonal experimental design based on previous studies.

Therefore, this study aims to explore the most suitable hormone combinations for mulberry shoot cutting based on existing research in order to breed more excellent mulberry seedlings in a short period of time, meet market production and market demand, and provide a theoretical basis and technical guidance for mulberry shoot cutting propagation. To this end, we conducted a green branch cutting experiment on Yueshenda 10. By applying orthogonal experimental design, we compared the effects of different influencing factors on the rooting characteristics of Yueshenda 10 during the tender branch cutting process.

2. Materials and Methods

2.1. Materials

The experiment was conducted in the third residential area of Henan Agricultural University in Zhengzhou City, Henan Province. Henan Province is the birthplace of silkworm production, which includes subtropical and warm temperate regions with an average annual temperature of 12~15 °C, annual rainfall of 600~1200 mm, and 2000~2600 h of sunshine per year. The region has a large temperature difference between day and night [28]. The third residential area of Henan Agricultural University is located at 113.22° E and 34.28° N and is equipped with a fully-illuminated automatic spray greenhouse with a propagation pool measuring 11 m in length, 6 m in width, and 0.4 m in depth. The pool is evenly divided into five small propagation pools, each with a length of 6 m, width of 2 m, and depth of 0.4 m.

The mulberry cuttings used for experiments were obtained from 3-year-old mulberry trees in the experimental mulberry orchard located in the third residential area of Henan Agricultural University. "Yueshenda 10" (*Morus atropurpurea* Roxb. 'Yueshenda 10') is a fruit mulberry variety selected by the Institute of Sericulture and Agroprocessing, Guangzhou Academy of Agricultural Sciences, Guangdong Province, China. This germplasm was introduced and preserved in the ex situ conservation bank of mulberry germplasm resources of Henan Agricultural University. The test material was selected from 3-year-old live seedlings of "Yueshenda 10" in the mulberry garden of the third living area of Henan Agricultural University.

The cuttings were made from semi-lignified branches, measuring 15 cm long with two to three semi-leaves (retaining half of the leaf blade), a flat cut on the upper end, and a 45° oblique cut on the lower end. The indole-3-acetic acid (IAA), indole-3-butyric acid (IBA), naphthaleneacetic acid (NAA), and ABT1 rooting powder used in this study were all purchased from Solabio Technology Co., Ltd. in Beijing, China. The growth factors used in this experiment were indole-3-butyric acid (IBA), indole-3-acetic acid (IAA), and naphthalene acetic acid (NAA) produced by Beijing Solarbio Science & Technology Co., Ltd., Beijing, China. Rooting powder (ABT1) was produced by the Forestry Research Institute of China Academy of Forestry Science. The propagation substrate was river sand purchased from the market.

2.2. Experimental Design

We designed an experiment to investigate the effects of plant growth hormones, the concentration, and the immersion time on the rooting of mulberry cuttings. This experiment adopts an orthogonal design with three factors: the exogenous hormone concentration type, exogenous hormone concentration, and soaking time, represented by A, B, and C. Each factor has four levels as shown in Table 1. For each factor, the tests were performed at four levels as shown in Table 1. The L16(4^3) orthogonal experimental design was employed (Table 2) with 17 treatments, including a control group (CK). Each treatment was replicated three times, and 150 cuttings were used in each repeat. Using the orthogonal experimental design of L16 (43) (Table 2), a total of 17 treatments were added with water treatment (CK). Each treatment was repeated three times, with 50 cuttings per repetition, resulting in a total of 2550 cuttings.

Table 1. Test factors and levels.

Experimental Level	Experimental Factors		
	Types of Hormones(A)	Concentration/mg/L(B)	Soaking time/Min(C)
1	NAA(A1)	200(B1)	10(C1)
2	IBA(A2)	500(B2)	30(C2)
3	IAA(A3)	800(B3)	60(C3)
4	ABT1(A4)	1000(B4)	120(C4)

Initially, the soil in each cutting pot was disinfected by spraying with a 1:400 dilution of carbendazim, turned over at least three times, and dried for 3 days. On the day of cutting, the cuttings were dipped in a 5% water solution of carbendazim for 10 s, while cuttings for the CK group were treated with water. The cuttings were then inserted into the pots according to the experimental design. After cutting, an automatic spray device was used to enhance the moisture and maintain the relative humidity of the air at 70% to 80%. Before cutting, according to a 1:400 dilution of carbendazim, spray application for each cutting pool was performed three times for soil disinfection, and the samples were dried for 3 days.

On the morning of the cutting day, cuttings were cut from a well-growing and pest-free mother tree and immediately treated.

Table 2. L16(4^3) three-factor four-level orthogonal experimental design.

Treatment Number	Treatment Combination	Types of Hormones(A)	Concentration/ mg/L(B)	Soaking Time/Min(C)
1	A1B1C1	NAA(A1)	200(B1)	10(C1)
2	A1B2C2	NAA(A1)	500(B2)	30(C2)
3	A1B3C3	NAA(A1)	800(B3)	60(C3)
4	A1B4C4	NAA(A1)	1000(B4)	120(C4)
5	A2B1C2	IBA(A2)	200(B1)	30(C2)
6	A2B2C1	IBA(A2)	500(B2)	10(C1)
7	A2B3C4	IBA(A2)	800(B3)	120(C4)
8	A2B4C3	IBA(A2)	1000(B4)	60(C3)
9	A3B1C3	IAA(A3)	200(B1)	60(C3)
10	A3B2C4	IAA(A3)	500(B2)	120(C4)
11	A3B3C1	IAA(A3)	800(B3)	10(C1)
12	A3B4C2	IAA(A3)	1000(B4)	30(C2)
13	A4B1C4	ABT1(A4)	200(B1)	120(C4)
14	A4B2C3	ABT1(A4)	500(B2)	60(C3)
15	A4B3C2	ABT1(A4)	800(B3)	30(C2)
16	A4B4C1	ABT1(A4)	1000(B4)	10(C1)
17	Control group (CK)	Water	0	10

First, we used 50% carbendazim water at 800 times liquid speed for 10 s. Secondly, we prepared solutions for orthogonal experiments. The method is as follows: We used a balance to weigh 500 mg of IAA. First, we dissolved 500 mg of IAA in a small amount of 75% ethanol, and then we added 1 L of water to prepare a 500 mg $\cdot$ L^{-1} IAA solution. We prepared IAA solutions of other concentrations and IBA, NAA, and ABT1 solutions in the same way. Finally, we placed the cuttings in the prepared plant growth regulator solution and slowly soaked them according to the experimental arrangement (Table 2). After cutting, the humidity was increased by automatic intermittent spraying facilities arranged in the shed to maintain the relative humidity of the air at 70–80%.

2.3. Measurement of Indicators

Three stems with similar stages of development were collected from each treatment group on days 1, 10, 20, 30, and 40 after cutting, and 2 cm of the bark was promptly sliced and blended with scissors. This process was repeated three times, and the samples were wrapped in tin foil, stored briefly in liquid nitrogen, and then stored at −80 °C until further measurement.

Analysis of rooting traits: at 50 days after cutting, several rooting traits, including the rooting rate, number of roots, average root number, average root length, longest root length, and rooting effect index [29], were evaluated for the 17 treatments. The formulae for these indicators are as follows:

Rooting rate = Number of cuttings rooted/Total number of cuttings × 100%

Average root number = Total root number of cuttings/Number of rooted cuttings

Average root length = Total root length of cuttings/Number of roots

Rooting effect index = (Average root length × Number of rooted cuttings)/Total number of cuttings.

2.4. Data Analysis

Data preprocessing and statistical analysis were conducted using Excel 2003. Range analysis was performed to preliminarily identify the optimal combination of treatments. Multiple-factor variance analysis was performed using SPSS 24.0 to determine the influence of each factor on the experimental results, and the least significant difference method (LSD) was employed to conduct multiple comparisons. The membership function method was applied to perform comprehensive evaluation of the rooting and propagation effects of each treatment combination. We drew the chart using Origin 2021.

The membership function calculation method was as follows: U(Xj) = (Xj − Xmin)/(Xmax − Xmin). The membership values of different indicators for each treatment were summed, and the average value was ranked. The larger the average value, the better the rooting effect.

3. Results and Analysis

3.1. Rooting Types and Process of Cutting Propagation

The rooting types of softwood cuttings were observed after planting, and the results showed that the main rooting type was callus formation (type II), followed by epidermal rooting (type I), and few roots were formed at the junction of callus and epidermis, which was classified as the mixed rooting (type III) (see Figure 1). The rooting process of cuttings was divided into five stages (see Figure 2): the callus formation stage (stage II) from 0 to 10 days after planting, the induction stage of root primordium (stage III) from 10 to 20 days, the expression and formation stage of adventitious roots (stage IV) from 20 to 30 days, and the elongation and development stage of adventitious roots (stage V) after 40 days. The control treatment resulted in the delayed formation of callus and adventitious root formation and had a lower rooting rate compared to the experimental treatments.

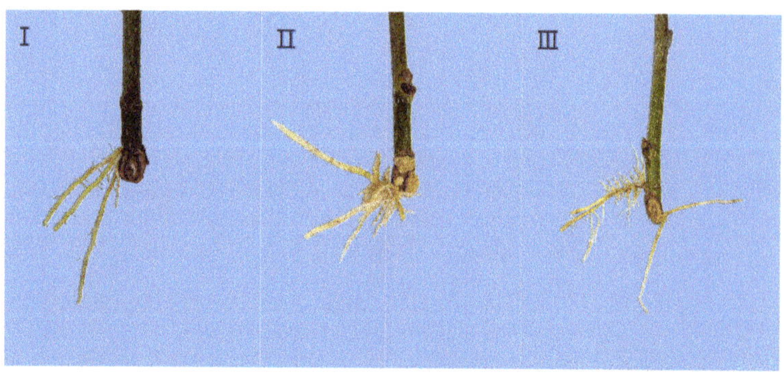

Figure 1. The rooting type of cuttings. (**I**) the root epidermis rooting, (**II**) the callus rooting, and (**III**) mixed rooting.

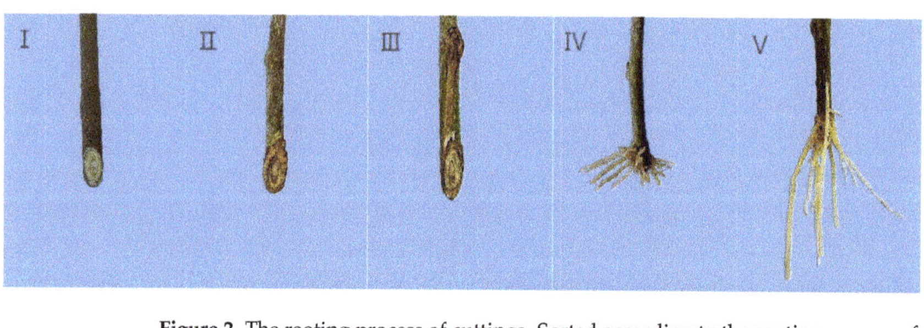

Figure 2. The rooting process of cuttings. Sorted according to the rooting process of cuttings (**I–V**).

3.2. Analysis of Primary and Secondary Effects of Different Factors

Figure 3 demonstrates that if the rooting rate is only taken as the primary indicator, the rooting rates for different types, concentrations, and treatment times of hormones were 29.11, 23.66, and 12.45, respectively. This demonstrates that the main factors affecting the softwood cutting rooting of mulberry were hormone type > hormone concentration > treatment time.

The values of K1, K2, and K3 indicated that the treatment combination with the best rooting promotion effect was the slow immersion of the cuttings in 800 mg/L ABT1 solution for 60 min (A4B3C3) with the highest rooting rate of 66.24%. When only the average root number was considered as the primary indicator, from the R-value of the range, it can be inferred that hormone concentration (3.30) > hormone type (2.38) > treatment time (1.57) had the least influence on the average root number, implying that hormone concentration had the strongest effect, and treatment time had the least effect.

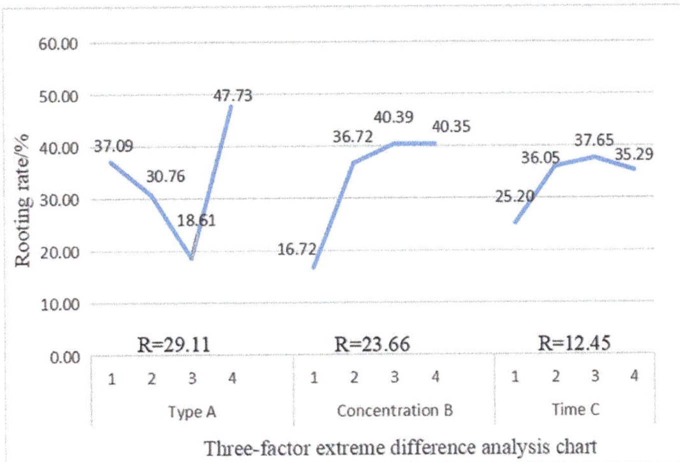

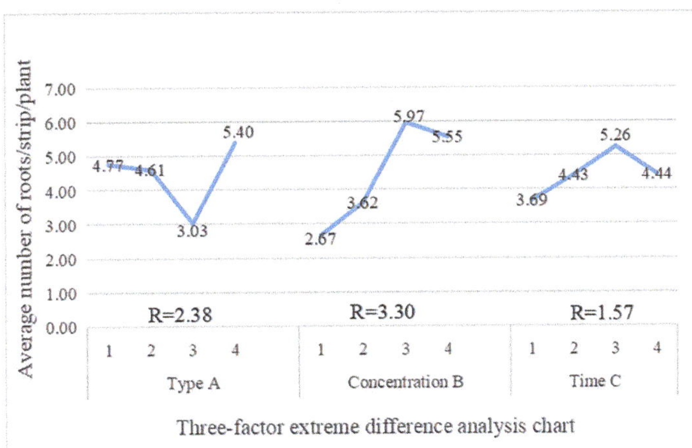

Figure 3. *Cont.*

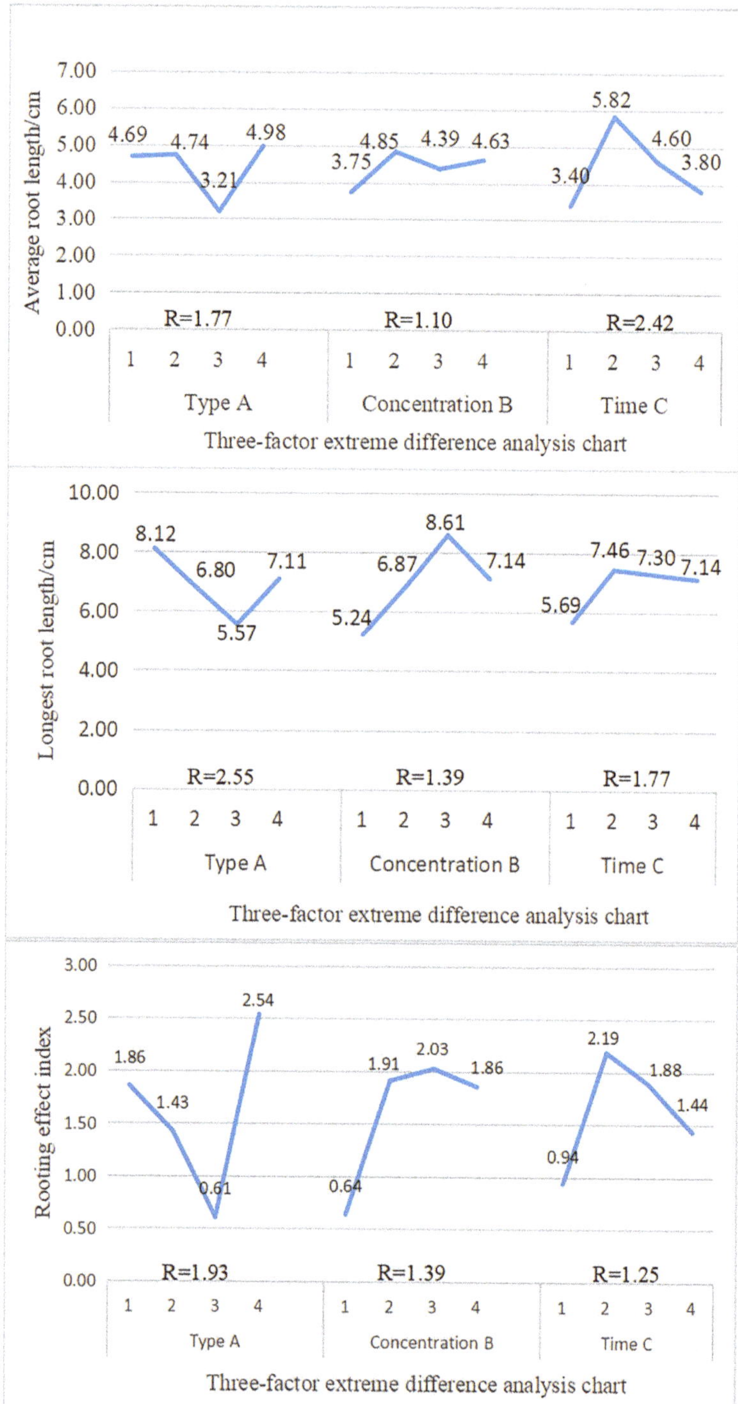

Figure 3. Range analysis for the influence of various test factors on the rooting index.

The K values indicate that the theoretical optimal combination is to soak the cuttings with 800 mg/L ABT1 for 60 min (A4B3C3). Considering only the indicator of average root length, the K1, K2, and K3 values of each factor indicated that the longest average root length of fruit mulberry could be obtained by soaking 500 mg/L ABT1 for 30 min (A4B2C2).

The R-values influencing the average root length of cuttings were treatment time (2.42) > hormone type (1.77) > hormone concentration (1.10), which suggests that treatment time was the most important factor influencing the average root length, whereas hormone concentration showed the smallest effect. If the longest root length was taken as the primary indicator, we observed that the order of influence on the longest root length of cuttings was hormone concentration > hormone type > treatment time, indicating that concentration (3.37) was the most important factor influencing the longest root length of softwood cuttings, whereas hormone type (2.55) and treatment time (1.77) had the least effects.

Based on the K1, K2, and K3 values of the three factors, the optimal combination of plant growth regulators was determined to be soaking cuttings with 800 mg/L NAA for 30 min (A1B3C2). Taking only the rooting effect index as the primary indicator, the R-value showed that the main factor influencing the rooting effect index of green shoot cuttings was the hormone type (1.93) > hormone concentration (1.39) > treatment time (1.25). In addition, analysis of the K1, K2, and K3 values of each factor revealed that A4B3C2 was the best combination for the mulberry softwood cutting rooting effect index, i.e., soaking in an 800 mg/L ABT1 solution for 30 min achieved the best rooting effect index of 4.23.

In terms of the rooting rate and average number of roots, soaking cuttings with 800 mg/L ABT1 for 30 min (A4B3C3) was the best combination. However, the other three theoretically optimal combinations were not observed in the experiment. It remains to be further tested whether the effectiveness of these combinations is influenced by the interaction of factor levels.

3.3. Effects of Hormone Types, Concentrations, and Soaking Time on Rooting Indicators of Yueshenda 10 Cuttings

Five indicators, including the rooting rate, average number of roots, average root length, longest root length, and rooting index, were used to evaluate the rooting effect of Yueshenda 10 green cuttings. The results were subjected to variance analysis (Table 3) and LSD multiple comparison tests (Figure 3) as shown in the table below.

Table 3. Correlation analysis between different hormone types, concentrations, and treatment times and changes in the rooting index of cuttings (* $p < 0.05$ and ** $p < 0.01$).

Source of Error	Rooting Rate		Average Number of Roots		Average Root Length		Longest Root Length		Rooting Index	
	F	*p*	F	*p*	F	*p*	F	*p*	F	*p*
A	6.245 *	0.016	3.730	0.060	1.977	0.165	0.016	0.901	4.555 *	0.039
B	24.746 **	0.000	59.980 **	0.000	7.157 *	0.011	32.649 **	0.000	12.957 **	0.001
C	3.070	0.087	3.034	0.089	0.978	0.328	11.926 **	0.001	1.035	0.315
A × B	0.000	0.983	0.829	0.368	3.415	0.072	0.996	0.324	0.165	0.687
A × C	0.172	0.680	0.016	0.900	1.079	0.305	6.899 *	0.012	0.004	0.950
B × C	5.706 *	0.021	3.272	0.076	1.523	0.024	0.011	0.917	8.492 **	0.006
A × B × C	0.000	0.997	0.264	0.610	7.060 *	0.011	2.222	0.143	1.496	0.228

Note: A: hormone type; B: Concentration; and C: Processing time. The "*" shows significant differences between each treatment while $p < 0.05$, and the "**" shows very significant differences between each treatment while $p < 0.01$. Data on the rooting rate were squared, and then the arcsine was subjected to analysis of variance.

3.3.1. Effects of Factors and Their Interactions on Rooting Indicators of Cuttings

Visual and straightforward representation of the experimental results can be achieved through range analysis; however, this does not convey information about the magnitude and precision of errors. Therefore, variance analysis should be performed [30]. The results of three-factor variance analysis (Table 3) showed that the concentration of growth regulators had the greatest impact on the rooting of cuttings ($p < 0.05$), except for the average root length. This concentration also had a significant effect on the rooting rate, average number of roots, longest root length, and rooting index ($p < 0.01$).

Hormone type caused a secondary effect and significantly affected the rooting rate and rooting index. The effect of the treatment time on rooting indicators was relatively small, only significantly affecting the longest root length, that is, hormone concentration > hormone type > treatment time. In terms of interaction among two factors, the interaction between the hormone concentration and treatment time had the greatest impact on rooting with significant and highly significant effects on the rooting rate and rooting index, respectively.

The interaction between hormone type and treatment time had a smaller effect, only significantly affecting the longest root length. The interaction effect between hormone type and concentration had no significant effect on any tested parameter. A comprehensive analysis concluded that the interaction effect between hormone concentration and treatment time was the main factor affecting the rooting of mulberry green cuttings.

3.3.2. Effects of Hormone Types on Rooting Index of Yueshenda 10 Cuttings

The data presented in Table 3 suggest that the significant probabilities (p values) of the rooting rate and rooting effect index among the five rooting indices of plant growth regulators were both below 0.05, indicating significant differences in the rooting rate and rooting effect index between hormone types. However, the differences in the other three rooting indices were not significant. Figure 4 also illustrates that the rooting rate, average root number, and rooting effect index were consistent among the four plant growth regulators and CK (distilled water) treatments.

The ranking of the five rooting indices from best to worst was as follows: ABT1, NAA, IBA, IAA, and CK. ABT1 treatment was significantly or extremely significantly different from other hormones, with ABT1 and NAA showing superior performance compared to IAA and CK. The rooting rate (47.73%), average root number (5.40 per plant), average root length (4.98 cm), and rooting effect index (2.54) under ABT1 treatment were the best, whereas those of NAA, IBA, and IAA were the worst.

3.3.3. Effects of Concentration on Rooting Index of Cuttings

As shown in Table 3, the F-values of the variance analysis results of different concentrations for each index are larger than the critical value, indicating significant differences. According to the multiple comparison results shown in Figure 4, the rooting rate, average number of roots, and longest root length exhibited similar responses to the four concentration levels and the CK (control) treatments. The optimal treatment was observed at a concentration of 800 mg/L, followed by concentrations of 1000 and 500 mg/L, whereas concentrations of 200 mg/L and CK were the least effective.

The longest average root length was observed at a concentration of 500 mg/L, and there were no significant differences between the four concentration levels and the CK treatment at the 0.05 and 0.01 significance levels, but all were significantly higher compared with the values for CK. The rooting rate and rooting effect index revealed were not significantly different among concentration levels 2, 3, and 4 but were significantly higher than levels 1 and the control treatment. The rooting rate (40.39%), average number of roots (5.97 per plant), longest root length (8.61 cm), and rooting effect index (2.03) were all optimal at a hormone concentration of 800 mg/L.

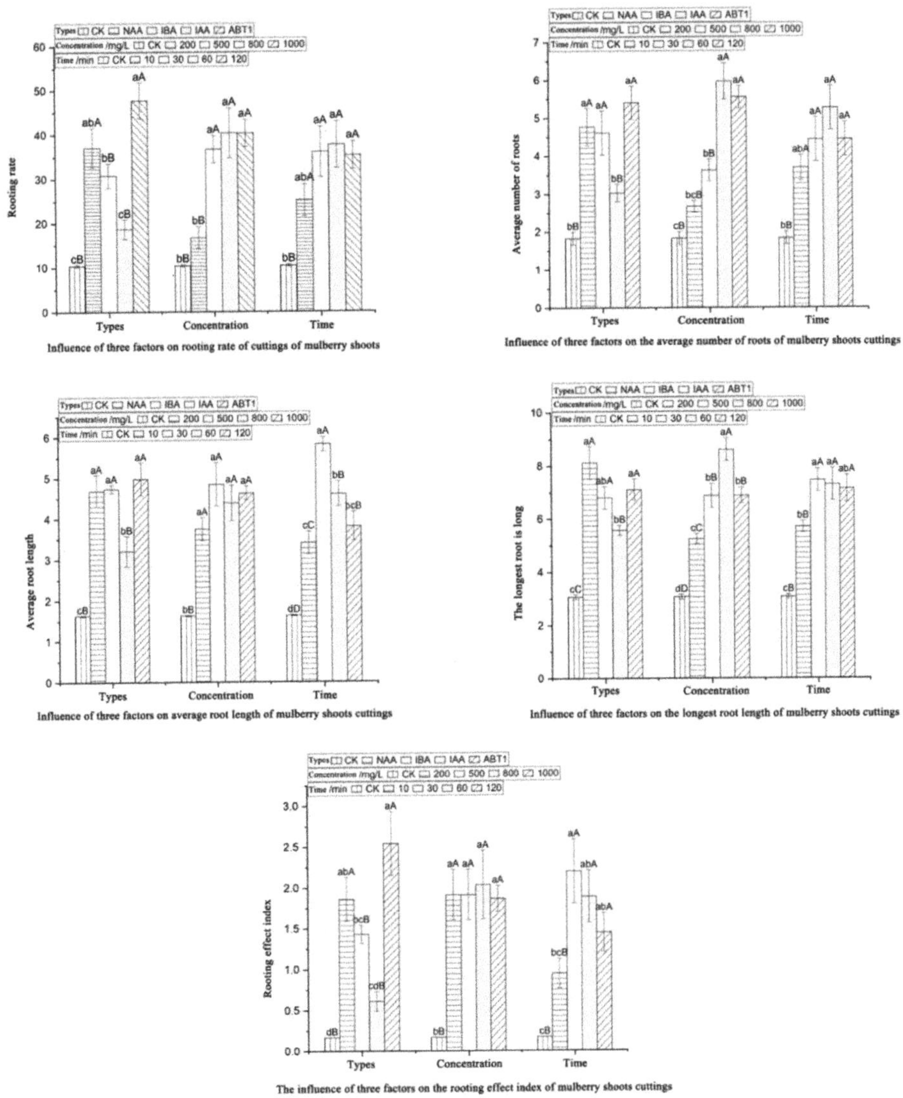

Figure 4. LSD multiple comparisons of the effects of the hormone type, hormone concentration, and time on the cutting rooting parameters of Yueshenda 10. Under each influencing factor, different uppercase letters indicate extremely significant differences between treatments ($p < 0.01$), while different lowercase letters indicate significant differences between treatments ($p < 0.05$).

3.3.4. Effects of Treatment Time on Rooting Indexes of Stem Cuttings

The rooting rate (37.65%) and mean root number (5.26 cm) were the highest after 60 min of treatment. At the three treatment durations of 30, 60, and 120 min, the above parameters were not significantly altered at the 0.05 and 0.01 levels but were significantly higher than the 10 min and water treatments with extremely significant differences compared to the water treatment. The mean root length, maximum root length, and rooting effectiveness index showed consistent results among the four treatment durations and the water treatment, with the rooting effectiveness rankings being 30 min, 60 min, 120 min, 10 min, and water treatment, in that order.

The mean root length (5.82 cm), maximum root length (7.46 cm), and rooting effectiveness index (2.19) were optimal at the 30 min treatment duration. Significant differences were recorded in the mean root length among different soaking times, with the 30 min treatment being extremely significant higher compared with the levels at the 60 and 120 min treatments, and levels 3 and 4 being extremely significantly higher than the 10 min and water treatments. The maximum root lengths of the 30 and 60 min treatments were not significantly different at the 0.05 and 0.01 levels but were significantly higher than the 120 min treatment and were extremely significantly higher relative to the other treatment levels.

The rooting effect index showed no significant difference between the 60 min and 120 min treatments ($p > 0.05$), but the effect between the 60 min and 120 min treatments and the 10 min and clear water treatments was extremely significant ($p < 0.01$).

3.4. Fuzzy Function Analysis of the Optimal Combination of Hormone Type, Concentration, and Soaking Time for the Rooting of Yueshenda 10 Stem Cuttings

The study discovered notable variations in the four rooting indicators across the 16 treatment combinations. The various rooting indicators did not exhibit identical levels of effectiveness within the same treatment combination. Therefore, using the membership function method, a comprehensive analysis was conducted on the five rooting indicators, including the rooting rate, average root number, average root length, longest root length, and rooting effect index, under the 16 treatment combinations. The ranking of each treatment U(Xj) is shown in Table 4.

Table 4. Effects of different factor combinations on indexes of mulberry softwood cutting and evaluation of the membership function.

Treatment Combination		Rooting Rate (%)	Average Root Number (Root/Plant)	Average Root Length/cm	Length of Longest Root/cm	Rooting Effect Index	Mean of Membership Function	Sequence
15	A4B3C2	66.240 ± 0.430	7.538 ± 0.024 a	6.378 ± 0.004 b	8.833 ± 0.088 c	4.225 ± 0.028 a	0.959	1
14	A4B2C3	52.953 ± 0.454	5.111 ± 0.016 e	6.230 ± 0.004 c	7.800 ± 0.100 e	3.299 ± 0.026 b	0.747	2
3	A1B3C3	47.290 ± 0.337	6.885 ± 0.015 b	4.371 ± 0.002 j	10.200 ± 0.058 a	2.067 ± 0.014 e	0.720	3
4	A1B4C4	52.780 ± 0.284	5.726 ± 0.014 d	5.089 ± 0.012 f	8.567 ± 0.088 d	2.686 ± 0.009 c	0.712	4
2	A1B2C2	36.400 ± 0.243	3.747 ± 0.018 g	6.444 ± 0.013 a	8.900 ± 0.058 c	2.346 ± 0.014 d	0.635	5
8	A2B4C3	42.300 ± 0.560	6.860 ± 0.016 b	4.463 ± 0.006 i	6.633 ± 0.033 f	1.888 ± 0.024 f	0.596	6
16	A4B4C1	41.653 ± 0.269	5.557 ± 0.030 d	3.792 ± 0.003 k	6.433 ± 0.033 g	1.579 ± 0.010 g	0.500	7
7	A2B3C4	32.303 ± 0.527	6.097 ± 0.050 c	4.530 ± 0.002 h	9.100 ± 0.115 b	1.464 ± 0.025 h	0.466	8
12	A3B4C2	24.667 ± 0.333	4.056 ± 0.056 f	5.180 ± 0.004 e	5.867 ± 0.033 i	1.278 ± 0.018 i	0.416	9
6	A2B2C1	31.550 ± 0.165	3.103 ± 0.052 i	4.680 ± 0.003 g	5.233 ± 0.067 k	1.476 ± 0.009 h	0.377	10
5	A2B1C2	16.910 ± 0.240	2.361 ± 0.073 kl	5.277 ± 0.006 d	6.233 ± 0.033 h	0.892 ± 0.012 k	0.325	11
13	A4B1C4	30.060 ± 0.394	3.411 ± 0.048 h	3.535 ± 0.022 l	5.367 ± 0.033 k	1.063 ± 0.012 j	0.319	12
11	A3B3C1	15.723 ± 0.182	3.361 ± 0.073 h	2.273 ± 0.002 o	6.300 ± 0.058 gh	0.357 ± 0.004 m	0.207	13
10	A3B2C4	25.997 ± 0.508	2.533 ± 0.033 jk	2.045 ± 0.003 p	5.533 ± 0.033 j	0.532 ± 0.011 l	0.191	14
1	A1B1C1	11.870 ± 0.406	2.722 ± 0.147 j	2.852 ± 0.008 n	4.800 ± 0.058 l	0.339 ± 0.011 m	0.152	15
9	A3B1C3	8.057 ± 0.471	2.167 ± 0.167 l	3.348 ± 0.004 m	4.567 ± 0.033 m	0.270 ± 0.015 n	0.125	16
17	CK	10.413 ± 2.309	1.833 ± 0.167 m	1.628 ± 0.015 q	3.067 ± 0.088 n	0.169 ± 0.003 o	0.008	17

Note: The data in the table are the average values of three repetitions ± the standard error; Different letters in the same column indicate significant differences between treatments ($p < 0.05$), while the same letter indicates no significant difference between them ($p > 0.05$).

The results showed that, among the top five treatments in terms of the overall cutting effect U (Xj), the main types of exogenous hormones were ABT1 and NAA, and the rooting rates of the top four combinations were also in the top four. It can be seen that hormone types have a significant impact on the rooting rate of cuttings, which is consistent with the results of the analysis of variance. The membership function value of the three-factor treatment combination 15 was the highest, i.e., the rooting effect of cuttings was optimal under the A4B3C2 treatment, which is consistent with the optimal combination analysis of the rooting effect index.

The rooting rate (66.24%), average root number (7.54), and rooting effect index (0.96) were ranked the best, whereas other indicators ranked were ranked among the top four. In other words, the best softwood cutting effect of Yueshenda 10 was achieved following treatment of ABT1 at a concentration of 800 mg/L and a treatment time of 30 min. The combination of soaking with 500 mg/L ABT1 for 60 min was the second best, with the rooting rate (52.95%), average root length (6.23 cm), and rooting effect index (3.30) being ranked among the top three.

Although the rooting rate (47.29%) of the combination of soaking with 800 mg/L NAA for 60 min was not ranked among the top three, its longest root length (10.30 cm) and average root number (6.89) both showed excellent performance, making it ideal for the cultivation of high-quality seedlings with developed root systems in areas with abundant cutting materials.

4. Discussion

Root cuttings can be divided into three types: epidermal rooting [31], callus rooting [32,33], and combined rooting, with the number and proportion of roots in each part being the main classification criterion. In this study, basal morphological analysis of rooting cuttings revealed that the majority of roots of the Yueshenda 10 tender shoot cuttings originated from callus tissue, whereas a small number of cuttings developed adventitious roots originating from the epidermis, and very few had roots originating from both sites.

This is consistent with the results reported by Shen et al. [34] for mulberry cutting seedlings, indicating that softwood cuttings of fruit mulberry belong to the callus rooting type, and that fruit mulberry is a difficult-to-root species. In addition, by observing the growth of cuttings' roots, it can be seen that the devel-opment of adventitious roots of mulberry branches went through five main stages. Next, the anatomical structure of roots is studied to confirm whether this judgment is sufficiently correct.

The process of cutting propagation is not only affected by the genetic characteristics, ecology, and biology but is also modulated by several external factors, such as light, air, temperature, and humidity [35]. Many studies have shown that exogenous plant hormones can promote the formation of adventitious roots in cuttings and can induce rooting in plants that are difficult to root. Different types of exogenous hormones promote the formation of adventitious root to different degrees [36–38]. In this study, the effects of different types of growth regulators, different concentrations, and treatment times, as well as their interactions on the rooting of softwood cuttings of *Morus alba* L., were comprehensively explored.

4.1. Effects of Plant Growth Regulators on Mulberry Cuttings

Currently, the hormones commonly used in production include ABT rooting powder, indole-3-acetic acid (IAA), indole-3-butyric acid (IBA), naphthylacetic acid (NAA), naphthaleneacetic acid amide, and other benzoxycarboxylic acid compounds. In this study, all plant growth regulators used promoted the rooting of softwood cuttings of mulberry.

This may be because exogenous IBA can potentially enhance the nutritional level of the rooting zone of trees, the content of endogenous hormones, and the activity of peroxidase, thus, providing suitable conditions for the formation of adventitious roots and increasing rooting rate and root growth [39–41]. ABT can not only potentiate the effects of exogenous hormones and rooting substances needed for cutting rooting but also can promote the synthesis of endogenous auxin, which accelerates the healing of the lower cutting surface and promotes rooting [42]. Some studies have also reported that NAA is a potential treatment for plants that show an impaired ability to root [43,44].

Significant research has shown that ABT No. 1 is suitable for plants with high economic value and that are difficult to root [45]. It has been found that ABT No. 1 is more effective than NAA and IBA in rooting studies of Catalpa bungei [46], wild European plum [47] and Ulmus macrocarpa [48]. There is also evidence that IAA is less effective than IBA and NAA in promoting cutting rooting [49,50]. Geng Wenjuan et al. [47] reported that the survival rate of cuttings treated with 800 mg/L ABT rooting powder was as high as 80.00%, and the rooting rate was as high as 76.67%.

In this study, NAA, IBA, IAA, and ABT1 were used to treat cuttings at four levels, and the results showed that ABT1 had the best seedling growth effect with the highest rooting rate, average number of roots, average root length, and rooting effect index. In comparison, NAA and IBA were less effective than ABT1, whereas IAA had the poorest effect. The effects of the four growth regulators on the seedling growth were consistent with those reported previously.

In addition, evidence from similar studies has shown that the effect of cutting rootings under mixed treatment with plant growth regulators is better than that of using a single growth regulator [22,51]. For example, when using a mixture of NAA and IBA to treat softwood cuttings, the mixed treatment showed better outcomes compared with NAA or IBA alone [52]. However, the combined use of growth regulators was not explored in this study and, thus, needs to be further investigated.

4.2. *Effects of Plant Growth Regulator Concentration on the Rooting of Mulberry Cuttings*

Plant growth regulator (PGR) concentration is one of the main factors affecting the rooting of plant cuttings. Appropriate concentrations can improve the rooting and root growth, while excessively high concentrations can damage the cutting tissues and impair the rooting process [53,54].

Plant growth regulator concentrations showed significant effects ($p < 0.05$) on the rooting of hairy rosemary [55] and hops [56] plug cuttings in rooting studies. Meng Haisan et al. [57] concluded that the effect of exogenous hormone concentrations on the rooting rate, mean root length, and mean root number of blueberry cuttings reached significant levels ($p < 0.05$). In addition, the main factor affecting the root length of baldcypress cuttings was also the hormone mass concentration [58]. Our results are consistent with those of previous studies, suggesting that the PGR concentration significantly affects all the tested indicators. Wang Bangqin et al. found that, with an ABT1 soaking time of 0.5 h, the rooting rate of cuttings increased with the PGR concentration, reaching a peak before decreasing [59].

Luo Xuemei et al. [60] and Xu Yin et al. [61] reported that, in softwood cutting experiments, ABT rooting powder had the best overall cutting performance at a concentration of 800 mg/L, which is consistent with the results of this study.

Furthermore, this study observed a relatively uniform pattern of response from the different indicators to varying levels of PGR concentration. As the concentration of PGR quality increased, the rooting rate, average number of roots, maximum root length, and rooting efficiency index all demonstrated an initial increase followed by a subsequent decrease. As the concentration gradually increased within the range of 0~200~500~800 mg/L, it began to decrease at 1000 mg/L. The optimal treatment was at 800 mg/L, and within a certain range, the higher the quality concentration, the better the rooting effect, but excessive PGR concentration can inhibit rooting.

In contrast, Rovier, V. et al. concluded, in a study on Brazilian native medicinal plants, that the application of high concentrations of IBA could better promote the root development of the spike with a maximum rooting rate of 79.17% [62]. The differences between the results of this study and theirs may be due to differences in the variety, genetics, and physiological conditions of the cuttings or may be related to the experimental design.

4.3. *Effects of Treatment Time on the Propagation of Mulberry Cuttings*

The soaking time of plant growth regulators can affect the propagation index of cuttings. Cheng et al. found that the rooting rate of softwood cuttings in maple was the highest when soaked in NAA or ABT for 30 min [63]. Chen et al. demonstrated that soaking time significantly affected the survival rate, total number, and length of roots in Betula albosinensis cuttings with the best results obtained after a 2 h soaking period [64].

In the study, soaking time only had a significant effect on the longest root length of tender mulberry cuttings, and it had no significant effects on the rooting rate, average number of roots, average root length, and rooting effectiveness index, which is inconsistent with the results by Chen et al. This discrepancy may be related to the plant's own genetic factors, physiological conditions, and environmental conditions during propagation.

The results of the orthogonal test conducted revealed that the best treatment approach for enhancing the rooting rate of delicate mulberry softwood cuttings was immersing them in 800 mg/L ABT1 for 30 min. However, this particular approach was not deemed the most effective for increasing the average or maximum length of the roots. In the actual reproduction production of "Yueshenda 10", different combinations of plant growth regulators can be selected based on the desired propagation index. For example, if a longer average root length is required, cuttings can be soaked in 500 mg/L NAA for 30 min, while if a longer longest root length is desired, cuttings can be soaked in 800 mg/L NAA for 1 h.

4.4. Effects of Interaction on the Propagation of Mulberry Cuttings

Numerous studies have shown that the hormone type, concentration, and time not only individually affect the rooting of cuttings but also have significant interactive effects on each other [65,66]. Quan et al. reported that the concentration and soaking time of IBA, as well as their interaction, significantly influenced the rooting rate, number of roots, root length, and rooting index of cuttings [67]. Shen et al. discovered that different concentrations and soaking times of NAA had significant effects on cutting rooting, and the interaction between different concentrations of NAA and soaking time also affected the rooting [68].

Moreover, the interaction between different concentrations of ABT1 and soaking time showed strong modulatory effects on the rooting of semi-lignified shoots in the current year. However, Jussara et al. found that the interaction between hormone types and treatment time did not significantly affect the rooting of cuttings [69]. In this study, we found that the interaction between hormone concentration and treatment time significantly and extremely significantly affected the rooting rate and rooting index of cuttings, respectively.

Yi et al. found that the interaction effect of the rooting hormone and soaking time had an extremely significant effect on the rooting of stem cuttings of Cyclocarya paliurus [70]; Hu et al. found that the interaction effects of hormone type × hormone concentration, hormone type × soaking time, and three-factor interaction significantly affected the rooting percentage and longest root length of Picea abies cuttings [71].

Our results show that the interaction effects of hormone type × treatment time mainly influenced the longest root length of the cuttings, while the three-factor interaction only had a significant effect on the average root length. Moreover, none of the interactive effects influenced the average number of roots. These results are consistent with the findings by Ou et al., who reported that the interaction effect between rooting promoters and concentration did not affect the rooting percentage, average number of roots, and total root length of cuttings [72], and this was different from the findings by Yi et al. [70].

5. Conclusions

This study indicates that the Yueshenda 10 mulberry cultivar belongs to the type of callus rootings, and its softwood cuttings undergo five developmental stages during the propagation process. Among the factors affecting the softwood cuttings, hormone concentration was the key factor, followed by hormone type, while soaking time had little effect. In addition, the optimal treatment combination for softwood cuttings (800 mg/L ABT1, soaking for 30 min, and using river sand as the rooting substrate) was determined, which achieved a rooting rate of 66.24% and effectively solved the practical problem of difficult rooting in this cultivar of mulberry. These treatment settings can, therefore, be applied in large-scale seedling cultivation.

Exogenous auxin treatment altered the growth level and interrelationships among the five rooting indicators during the rooting process of Yueshenda 10 softwood cuttings, which indicated specific changes and was closely related to the rooting initiation and development. However, the rooting of softwood cuttings is an extremely complex physiological process, and subsequent research based on this study will focus on the composite physiological and biochemical characteristics and genetic traits of mulberry seedlings to further reveal the rooting mechanism of Yueshenda 10 softwood cuttings and clarify the mechanisms underlying the mulberry rooting process.

The exogenous auxin treatment changed the growth level and relationship of five rooting indicators in the rooting process of "Yueshendashi" cuttings, which showed a certain regularity in the rooting process and was closely related to the occurrence and development of cuttings rooting. This experiment found that hormone concentration was the key factor affecting the rooting changes of cuttings, followed by hormone type, and treatment time had little effect.

In addition, the optimal treatment combination for softwood cutting was determined to be soaking the cuttings in 800 mg/L ABT1 for 30 min using river sand as the rooting medium, and the highest rooting rate reached 66.24%. This treatment effectively solves the problem of the actual rooting difficulties for this variety of mulberry tree and can be considered for large-scale seedling cultivation. In the future, we will further study the molecular mechanism of adventitious roots of mulberry in terms of auxin regulation.

Author Contributions: Conceptualization—methodology, J.Q. and J.S.; software—validation, H.C.; search. H.C. and T.W.; writing—original draft software, J.S.; validation, H.L. and T.W.; formal analysis, H.C. and H.L.; research work, J.S.; data collection, J.Q. and H.C.; original manuscript drafting, J.S.; review preparation and editing, J.S.; guidance, J.Q. and H.B.; administration, H.B.; funding procurement, H.B. and J.Q. All authors have read and agreed to the published version of the manuscript.

Funding: This work has received funding from the National Natural Science Foundation of China and the Forestry Bureau of Henan Province, China.

Institutional Review Board Statement: Not applicable.

Informed Consent Statement: Not applicable.

Data Availability Statement: The basic data for this article can be found in the article. However, some data is currently not shared and is also part of ongoing research. If necessary, it can be obtained from the corresponding author upon reasonable request.

Conflicts of Interest: The authors declare no conflict of interest.

References

1. Yongcheol, L.; Keum, T.H. Changes in physicochemical properties of mulberry fruits (*Morus alba* L.) during ripening. *Sci. Hortic.* **2017**, *217*, 189–196. [CrossRef]
2. Palli, j.; Baliva, M.; Biondi, F.; Calcagnile, L.; Cerbino, D.; D'Elia, M.; Piovesan, G. The Longevity of Fruit Trees in Basilicata (Southern Italy): Implications for Agricultural Biodiversity Conservation. *Land* **2023**, *12*, 550. [CrossRef]
3. Sourati, R.; Sharifi, P.; Poorghasemi, M.; Alves, V.E.; Seidavi, A.; Anjum, N.A.; Sofo, A. Effects of Naphthaleneacetic Acid, Indole-3-Butyric Acid and Zinc Sulfate on the Rooting and Growth of Mulberry Cuttings. *Int. J. Plant Biol.* **2022**, *13*, 21. [CrossRef]
4. Koyuncu, F.; Senel, E. Rooting of black mulberry (*Morus* nigra) hardwood cuttings. *J. Fruit Ornam. Plant Res.* **2003**, *11*, 53–57.
5. Mallick, P.; Ghosh, S.; Chattaraj, S.; Sikdar, S.R. Isolation of mesophyll protoplast from Indian mulberry (*Morus alba* L.) cv. S1635. *J. Environ. Sci.* **2016**, *13*, 217–222.
6. Koyuncu, F.; Balta, F. Adventitious root formation in leaf-bud cuttings of tea (*Camellia sinensis* L.). *Pak. J. Bot.* **2004**, *36*, 763–768.
7. Romero, J.L. A review of propagation programs for Gmelina arborea. *New For.* **2004**, *28*, 245–254. [CrossRef]
8. Sharma, V. Aeroponics for adventitious rhizogenesis in evergreen haloxeric tree *Tamarix aphylla* (L.) Karst.: Influence of exogenous auxins and cutting type. *Physiol. Mol. Biol. Plants* **2018**, *24*, 167–174. [CrossRef]
9. Mehri, H.; Mhanna, K.; Soltane, A.; Awad, E.; Rouili., F.; Abdelkarim., M. Performance of Olive Cuttings (*Olea europaea* L.) of Different Cultivars Growing in the Agro-climatic Conditions of Al-Jouf (Saudi Arabia). *Am. J. Plant Physiol.* **2013**, *8*, 41–49.
10. Lakehal, A.; Bellini, C. Control of adventitious root formation: Insights into synergistic and antagonistic hormonal interactions. *Physiol. Plant* **2018**, *165*, 90–100. [CrossRef]

11. Shen, N.; Chen, L.D.; Liu, K.; Zhou, G.L.; Chen, M.Y.; Pi, F.J.; Zhao, D.G.; Zhao, Y.C. Effect of auxin on rooting of tender branch cuttings of tea strain 'Zikui'. *Mol. Plant Breed.* **2022**, 1–22.
12. Sunil, P.; Verma, R.C. Vegetative propagation of *Dalbergia sissoo* Roxb. using softwood and hardwood stem cuttings. *J. Arid Env.* **1996**, *34*, 235–245.
13. Rakeshwar, S.R.; Kamal, K.S. Effect of cutting diameter and hormonal application on the propagation of *Ficus roxburghii* Wall. through branch cuttings. *Ann. For. Sci.* **2012**, *55*, 69–84.
14. Laubscher, C.P.; Ndakidemi, P.A. Rooting success using IBA auxin on endangered *Leucadendron laxum* PROTEACEAE in different rooting mediums. *Afr. J. Biomed. Res.* **2008**, *7*, 3437–3442.
15. Cooper, W. Transport of root-forming hormone in woody cuttings. *Plant Physiol.* **1936**, *11*, 779–793. [CrossRef] [PubMed]
16. Franzon, R.C.; Castro, C.M.; Raseira, M.D.C.B. Variabilidade genética em populações de pitangueira oriundas de autopolinização e polinização livre, acessada por AFLP Genetic variability in surinam cherry populations originated from self-pollination and cross pollination, estimated by AFLP. *Rev. Bras. Frutic.* **2010**, *32*, 240–250. [CrossRef]
17. Cox, D.A. Hartmann and Kester's Plant Propagation Principles and Practices, 9th Edition. *HortScience* **2018**, *53*, 741. [CrossRef]
18. Kentelky, E.; Jucan, D.; Cantor, M.; Szekely, V.Z. Efficacy of Different Concentrations of NAA on Selected Ornamental Woody Shrubs Cuttings. *Horticulture* **2021**, *7*, 464. [CrossRef]
19. Dave, A.; Vaistij, F.E.; Gilday, A.D.; Penfeld, S.D.; Graham, I.A. Regulation of *Arabidopsis thaliana* seed dormancy and germination by 12-oxo-phytodienoic acid. *J. Exp. Bot.* **2016**, *67*, 2277–2284. [CrossRef] [PubMed]
20. Kohanová, J.; Martinka, M.; Vaculík, M.; White, P.J.; Hauser, M.; Lux, A. Root hair abundance impacts cadmium accumulation in *Arabidopsis thaliana* shoots. *Ann. Bot.* **2018**, *122*, 903–914. [CrossRef]
21. Huang, D.L.; Feng, J.L. Overview of the application of plant growth regulators on mulberry trees. *Guangxi Seric.* **1999**, *36*, 1–5.
22. Cui, Q.Y.; Qiu, C.Y.; Lin, Q.; Zhang, C.H.; Zhu, G.H.; Shi, H.Y.; Chen, F.; Zhu, F.R. Research progress in efficient mulberry cutting seedling technology. *Guangxi Seric.* **2017**, *53*, 36–41.
23. Zhang, Y.J.; Zhang, X.J.; Zhang, G.Z.; Li, D.D.; Li, R.J. Research on hardwood cutting technology of Longsang. *Prac. For. Tech.* **2012**, *27*, 27–29. [CrossRef]
24. Liu, Y.G.; Du, W.; Zhou, C.T.; Huang, P. Effect of different concentrations of NAA on rooting of green branch cuttings of mulberry trees Yunsang 2 and Yunguo 1. *Modern Agric. Sci. Tech.* **2015**, *23*, 281+285.
25. Lin, Q.; Chen, X.Q.; Mo, X.H.; Yang, Q.B. Comparative test on survival rate of mulberry hardwood cuttings treated with several plant hormones. *Guangxi Seric.* **2002**, *39*, 5–7.
26. Du, W.; Cheng, J.L. Effect of different spikes on the rooting effect of mulberry green cuttings. *China Seric.* **2014**, *35*, 11–14. [CrossRef]
27. Deng, G.X.; Lei, W.; Li, J.; Jiang, Y.B.; Wan, R.W. Study on the influence of ABT1 rooting powder and indole-3-acetic acid on the survival rate of Simao pine cuttings. *West. For. Sci.* **2016**, *35*, 73–77. [CrossRef]
28. Zhang, J. Meteorological characteristics and mulberry cultivation in Henan. *Agric. Henan* **2019**, *522*, 14. [CrossRef]
29. Ji, K.; Wang, Z.R.; Chen, T.H.; Wang, M.X. A study on the variation of rooting ability of *Pinus massoniana* cuttings. *J. Nanjing For. Univ.* **1998**, *22*, 69–73.
30. Luo, M.F. Application of experimental design in forestry. *Henan Agric. For. Sci. Tech.* **1984**, *2*, 25–27.
31. Cao, X.; Du, W.; Shang, C.Q.; Shen, Q.D.; Liu, L.; Cheng, J.L. Comparative transcriptome reveals circadian and hormonal control of adventitious rooting in mulberry hardwood cuttings. *Acta. Physiol. Plant* **2018**, *40*, 197. [CrossRef]
32. Zem, L.M.; Weiser, A.H.; Zuffellato-Ribas, K.C.; Radomski, M.I. Herbaceous and semi-hardwood stem cuttings of *Drimys brasiliensis*. *Revista Cienc. Agron.* **2015**, *46*, 396–403.
33. Vieira, L.M.; Kruchelski, S.; Gomes, E.N.; Zuffellato-Ribas, K.C. Indolebutyric acid on boxwood propagation by stem cuttings. *Ornam. Hortic.* **2018**, *24*, 347–352. [CrossRef]
34. Shen, Z.X. Rooting physiology of mulberry cuttings. *Plant Physiol. Commun.* **1997**, *6*, 485–486. [CrossRef]
35. Fukuda, Y.; Hirao, T.; Mishima, K.; Ohira, M.; Hiraoka, Y.; Takahashi, M.; Watanabe, A. Transcriptome dynamics of rooting zone and aboveground parts of cuttings during adventitious root formation in *Cryptomeria japonica* D. Don. *BMC Plant Biol.* **2018**, *18*, 201. [CrossRef] [PubMed]
36. Zhang, P.J.; Gao, L.Q.; Shang, X.H. Effects of different hormone types, concentrations and soaking time on rooting of Jinputao cuttings. *J. Trop. For. Sci.* **2020**, *41*, 1084–1091.
37. Wiesman, Z.; Lavee, S. Enhancement of IBA stimulatory effect on rooting of olive cultivar stem cuttings. *Sci. Hortic.* **1995**, *62*, 189–198. [CrossRef]
38. Uwe, D.; Philipp, F.; Hajirezaei, M.R. Plant Hormone Homeostasis, Signaling, and Function during Adventitious Root Formation in Cuttings. *Front. Plant Sci.* **2016**, *7*, 381. [CrossRef]
39. Pandey, L.; Verma, R.S.; Shukla, K.K.; Shukla, J.K.; Pathak, S. Effect of IBA and NAA on Rooting of Stem Cutting in Dragon Fruit [*Hylocereus undatus* (Haworth) Britton & Rose]. *J. Exp. Agric. Int.* **2022**, *44*, 1–6. [CrossRef]
40. Azad, M.S.; Alam, M.J.; Mollick, A.S.; Khan, M.N. Rooting of cuttings of the wild Indian almond tree (*Sterculia foetida*) enhanced by the application of indole-3-butyric acid (IBA) under leafy and non-leafy conditions. *Rhizosphere* **2018**, *5*, 8–15. [CrossRef]
41. Muttaleb, Q.A.; Abdullah, T.L.; Rashid, A.A.; Hassan, S.A. Rooting of Stem Cuttings with Different Indole 3 Butyric Acid (IBA) Treatments and Development of Micropropagation Protocol for *Piper betle* L. Node Culture. *Am. J. Plant Sci.* **2017**, *8*, 3084–3100. [CrossRef]

42. Paradikovi, N.; Zeljkovi, S.; Tkalec, M.; Vinkovi, T.; Mari, M. Influence of rooting powder on propagation of sage (*salvia officinalis* L.) and rosemary (*rosmarinus officinalis* L.) with green cuttings. *Poljoprivreda* **2013**, *19*, 10–15.
43. Nakata, P.A. Advances in our understanding of calcium oxalate crystal formation and function in plants. *Plant Sci.* **2003**, *164*, 901–909. [CrossRef]
44. Wagner, C.; Bryan, W.L.; Berry, R.E.; Knight, R. Carambola Selection for Commercial Production. *Proc. Fla. State Hortic. Soc.* **1975**, *88*, 466–469.
45. Zheng, K.; Lang, N.J.; Cao, F.L.; He, L.P.; Chen, P.; Peng, M.J. Analysis of Cutting Technology. *Guizhou Agric. Sci.* **2009**, *37*, 195–199.
46. Zhou, Y.C.; Zhong, Q.P.; Li, Q.P.; Liu, W.F.; Zou, Y.L.; Tian, F.; Ge, X.N.; Cao, L.Q. Study on the cutting propagation and rooting mechanism of half lignified spring shoots of tung trees. *J. Central South For. Univ. Sci. Tech.* **2020**, *40*, 25–36. [CrossRef]
47. Geng, W.J.; Jie, G.; Ouyang, L.T.; Jing, J.Y.; Ma, B.Q. Effects of different substrates and hormones on the green branch cutting propagation of wild European plum. *Northeast Agric. Sci.* **2021**, *46*, 76–81. [CrossRef]
48. Song, B.Y.; Kang, Y.X.; Shi, J.D. Effects of plant growth regulators on rooting of *Ulmus elongatus* by cutting. *Shanxi For. Sci. Tech.* **2022**, *50*, 16–19+22.
49. Vigya, K.; Anitha, K.; Latha, R. Effect of auxins on adventitious rooting from stem cuttings of candidate plus tree *Pongamia pinnata* (L.), a potential biodiesel plant. *Trees* **2009**, *23*, 597–604. [CrossRef]
50. Sanjay, S.; Pramod, K.; Ansari, S.A. A simple method for large-scale propagation of *Dendrocalamus asper*. *Sci. Hortic.* **2003**, *100*, 251–255. [CrossRef]
51. Nair, A.; Zhang, D.L.; Smagula, J.; Hu, D.Y. Rooting and Overwintering Stem Cuttings of *Stewartia pseudocamellia* Maxim. Relevant to Hormone, Media, and Temperature. *HortScience* **2008**, *43*, 2124–2128. [CrossRef]
52. Yusnita, Y.; Jamaludin, J.; Agustiansyah, A.; Dwi, H. A Combination of IBA and NAA Resulted in Better Rooting and Shoot Sprouting than Single Auxin on Malay Apple *Syzygium malaccense* L Merr Perry Stem Cuttings. *AGRIVITA J. Agric. Sci.* **2017**, *40*, 80–90. [CrossRef]
53. Laubscher, C.P.; Ndakidemi, P.A. Rooting response under shade using IBA growth regulators and different growth mediums on *Leucadendron laxum* (Proteaceae)–A commercial cut flower. *Afr. J. Agric. Res.* **2008**, *3*, 740–746.
54. Tien, L.H.; Chac, L.D.; Oanh, L.; Ly, P.T.; Bao, T.B. Effect of auxins (iaa, iba and naa) on clonal propagation of solanum procumbens stem cuttings. *Plant Cell Biotechnology Mol. Biol.* **2020**, *21*, 113–120.
55. Lamia, V.M.; Rana, V.K.; Mohammad, B.H.; Elaheh, K.; Mohammad, A.A. Auxin concentration and sampling time affect rooting of *Chrysanthemum morifolium* L. and *Rosmarinus officinalis* L. *Azarian J. Agric.* **2016**, 3.
56. Mehri, S.; Vahid, R.S. Effects of media and indole butyric acid (IBA) concentrations on hopbush (*Dodoneae viscosa* L.) cuttings in green house. *Ann. For. Res.* **2012**, *55*, 61–68. [CrossRef]
57. An, H.S.; Meng, J.J.; Xu, F.J.; Jiang, S.; Wang, X.Q.; Shi, C.H.; Zhou, B.Q.; Luo, J.; Zhang, X.Y. Rooting Ability of Hardwood Cuttings in Highbush Blueberry (*Vaccinium corymbosum* L.) Using Different Indole-butyric Acid Concentrations. *HortScience* **2018**, *54*, 194–199. [CrossRef]
58. King, A.R.; Arnold, M.A.; Welsh, D.F.; Watson, W.T. Substrates, wounding, and growth regulator concentrations alter adventitious rooting of baldcypress cuttings. *HortScience* **2011**, *46*, 1387–1393. [CrossRef]
59. Wang, B.Q.; Yu, W.G.; Meng, Q.W.; Yang, Z.X.; Dou, L.C.; Song, X.Q.; Yu, X.D. Effect of plant growth regulators on the rooting of rattan oil star cuttings. *J. Tropical Bio.* **2016**, *7*, 48–52. [CrossRef]
60. Luo, X.M.; Xie, S.L.; Ma, J.H.; Shen, X.L.; Wang, X.H.; Ding, L.M. The technology of shoot cutting propagation of three *Acer palmatum Thunb* varieties. *Mol. Plant Breed.* **2019**, *17*, 7459–7466. [CrossRef]
61. Xu, Y.; Yuan, S.; Lu, Y.F.; Feng, Z.J. Effect of plant growth regulators on rooting of tender branch cuttings of *Phoebe breviflora*. *Zhejiang For. Sci. Tech.* **2017**, *37*, 69–73.
62. Verdi, R.; Bettoni, J.C.; Werner, S.S.; Boff, M.I.C.; Boff, P. Effects of the phenological stage, type of cutting and plant growth regulators on the propagation by stem cutting of *Poiretia latifolia Vogel*, a brazilian native medicinal plant. *Rev. Colomb. Cienc. Hortic.* **2020**, *14*, 267–274. [CrossRef]
63. Cheng, T.L.; Shi, J.S. China Maple Shoot Cutting Propagation Technology. *For. Sci. Tech. Dev.* **2003**, *17*, 36–37.
64. Chen, L.Y.; Qi, S.A.; Wang, K.F.; Yan, L.P.; Ren, F.; Wang, Y.H.; Liu, C.L. Influences of hormone types, concentrations and soaking time on cutting propagation of *Zelkova schneideriana*. *Shandong Agric. Sci.* **2018**, *50*, 72–76. [CrossRef]
65. Torkashvand, A.M.; Shadparvar, V. Rooting in hibiscus rosa-sinensis (yellow double hybrid) by indole butyric acid and rooting substrates. *OMICS Int.* **2012**, *2*, 194–197.
66. Swamy, S.L.; Puri, S.; Singh, A.K. Effect of auxins (IBA and NAA) and season on rooting of juvenile and mature hardwood cuttings of *Robinia pseudoacacia* and *Grewia optiva*. *New For.* **2002**, *23*, 143–157. [CrossRef]
67. Quan, W.Y.; Wang, H.Z.; Chen, H.G.; Song, M.S.; Zhai, Y.S.; Du, T. Effect of IBA on cutting rooting of *Lonicera japonica* Thunb. *Agric. Tech.* **2020**, *40*, 39–42. [CrossRef]
68. Shen, Z.; Li, T.H.; Wen, S.Z.; He, G.X.; Huang, S.H.; Tang, A.G. The influence of different factors on the rooting of tender branch cuttings of *Phoebe Minnan*. *J. Central South For. Univ. Sci. Tech.* **2013**, *33*, 63–67. [CrossRef]
69. Jussara, C.; Rejane, M.; Gerciane, C.; Silvanda, S.; Walter, E.; Carlos, E. Rooting of Herbaceous and Semihardwood Cuttings of Guava cv. Século XXI, under Varying Concentrations of *Indolebutyric Acid*. *J. Exp. Agric. Int.* **2019**, *33*, 1–9. [CrossRef]
70. Yi, H.L.; Wang, Y.; Xu, Y.; Luo, J.; Zhang, Q.M.; Wu, P.G.; Ren, H. Cutting propagation experiment of *Apterosperma oblata*. *Ecol. Sci.* **2016**, *35*, 184–188. [CrossRef]

71. Hu, M.H.; Ouyang, F.Q.; Jia, Z.R.; Wang, J.H.; Zhang, S.Z.; Ma, J.W. Study on influencing factors of European spruce cutting rooting and selection of excellent individual trees with rooting ability. *For. Sci.* **2014**, *50*, 42–49.
72. Ou, R.M.; Li, Z.H.; Liu, Y.Q.; Xiao, X.C.; Huang, L.Q.; Zhang, B.; Jiang, L.J. Effect of different rooting promoters on rooting of chestnut like cuttings. *J. Cent. South Univ.* **2011**, *31*, 139–143. [CrossRef]

life

MDPI

Article

Generation of Asynaptic Mutants in Potato by Disrupting *StDMC1* Gene Using RNA Interference Approach

Ashwani Kumar [1,2], Sundaresha Siddappa [1], Vinay Bhardwaj [1], Dalamu [1], Baljeet Singh [1,2], Neha Sharma [1], Bhawna Dipta [1], Vinod Kumar [1], Umesh Goutam [2] and Salej Sood [1,*]

1 ICAR-Central Potato Research Institute, Shimla 171001, India
2 Department of Bioengineering and Biosciences, Lovely Professional University, Phagwara 144411, India
* Correspondence: salej.sood@icar.gov.in or salej1plp@gmail.com

Abstract: Fixing the genomic composition and multiplication through true potato seed (TPS) is an important challenge in autotetraploid potato. Disrupted meiotic cDNA (DMC1) is a meiotic gene that plays a central role in DNA recombination through crossing over in meiosis. Using the *Arabidopsis* DMC1 (*AtDMC1*) gene sequence, we retrieved *Solanum tuberosum* DMC1(*StDMC1*) from the diploid potato genome, and subsequently, sense and antisense regions of the *StDMC1* gene were amplified in potato cv. Kufri Jyoti. The sense and antisense fragments were confirmed by Sanger-sequencing and cloned in the pRI101 vector. *Agrobacterium*-mediated transformation of the RNAi construct resulted in 44% transformation efficiency, and a total of 137 mutant lines were obtained. These mutant lines were further validated through pollen viability testing, and selected lines were used for gene expression analysis. The acetocarmine-based pollen staining showed reduced pollen viability ranging from 14 to 21% in four DMC1 mutant lines (DMC4-37, DMC4-41, DMC6-20, and DMC6-21), as compared to the Kufri Jyoti control plants, which on average exhibited 78% pollen viability. The phenotypic data was supported by the reduced expression of the *StDMC1* gene in these four mutant lines compared to the control Kufri Jyoti. The results confirmed the generation of *StDMC1* knockdown lines. This is the first report of *StDMC1* mutant line generation in tetraploid potatoes and will be a step forward in generating non-recombinant mutants through sexual reproduction in potatoes.

Keywords: Asynapsis; cross-over; *StDMC1*; meiosis; recombination; RNAi

Citation: Kumar, A.; Siddappa, S.; Bhardwaj, V.; Dalamu; Singh, B.; Sharma, N.; Dipta, B.; Kumar, V.; Goutam, U.; Sood, S. Generation of Asynaptic Mutants in Potato by Disrupting *StDMC1* Gene Using RNA Interference Approach. *Life* **2023**, *13*, 174. https://doi.org/10.3390/life13010174

Academic Editors: Sen Meng and Jie Luo

Received: 31 October 2022
Revised: 24 November 2022
Accepted: 19 December 2022
Published: 6 January 2023

1. Introduction

The potato (*Solanum tuberosum* L.) is the world's third most important food crop after wheat and rice [1] and is grown in all major countries [2]. Most of the cultivated potato varieties are autopolyploid and extremely heterozygous [3]. In potato breeding, heterozygous varieties are crossed with donor heterozygous genotypes to pass on desirable features, and the desired genotypes are chosen and released as new varieties in the F1 generation [4]. The hybrids cannot be recreated in the form of true potato seed (TPS) by crossing the same parental lines since each TPS in the F1 generation is genetically unique in potato [4]. Potato varieties are maintained in vitro and propagated clonally [5]. This is a significant distinction between potato breeding and that of key food crops such as rice, wheat, and maize, where homozygosity is first achieved before new varieties are released as open-pollinated homozygous lines or hybrid varieties. A constant influx of new genes and allelic diversity into the *S. tuberosum* gene pool is required to improve potato genetics for diverse economic traits. However, the introgression of new genes changes the genetic background of a genotype or a variety in potatoes [6], which has also been observed in other vegetables and crop wild relatives (CWRs) [7].

Potato seed production currently faces considerable challenges, such as low multiplication rates, high storage and transportation costs, virus carry-over, and performance reductions over generations [8]. Heterozygous genotypes cannot be sexually reproduced to form botanical seeds or TPS in light of the fact that parental chromosomes will recombine to produce different combinations and the same genotype cannot be reconstituted [9]. The favorable allele combinations are lost in the sexual cycle due to the segregation of traits. This is the reason that potatoes are reproduced clonally to maintain the favorable allelic combinations of genotypes/varieties [10]. Although apomixes is considered a solution to fix heterozygotes, it has not been applied in practical plant breeding in any crop [11]. Thus, fixing the genomic composition and multiplication through TPS is an important challenge in potato breeding [12]. An alternate method is the suppression of recombination and identification of asynaptic mutant lines, which could possibly recreate the parental type gametes and true-to-type botanical seeds similar in genetic constitution to starting clone/hybrid. The development of haploids and doubled haploids from asynaptic gametes is further required to create a set of parental lines, which, upon selection and hybridization, will recreate the original cultivar.

Homologous recombination (HR) is required for preserving genomic integrity across all domains of life, and it is also required for the repair of planned double-strand breaks (DSBs) during meiosis [13]. Meiotic HR results in chromosome segregation because of crossovers, while the purpose of mitotic HR is to repair DNA damage without crossovers. In mitotic cells, crossovers are less desirable since they can result in loss of homozygosity and other chromosomal rearrangements [14]. Crossovers, on the other hand, are preferred during meiosis because they are required for chromosome segregation and serve to increase genetic variation among offspring. Meiosis is an important step in the sexual reproduction cycle. It is an atomic division in which the number of chromosomes is reduced to half, which is essential for the arrangement of haploid cells in living organisms that reproduce sexually [15].

Meiotic recombination protein disrupted meiotic cDNA (DMC1) is a homologue of the bacterial strand trade protein RecA. DMC1 assumes the focal job in homologous recombination in meiosis by amassing at the destination. The protein encoded by this gene is basic for meiotic HR. Hereditary recombination in meiosis assumes a significant role in creating an assorted variety of hereditary data and encourages the reductional isolation of chromosomes that must happen in the development of gametes during sexual multiplication.

Meiotic recombination requires two types of RecA-like proteins, DMC1 and RAD51, that play an important role in HR. RecA protein is essential for the repair of DSBs, while both Rad51 and DMC1 are essential to generate a crossover between homologous chromosomes, which confirms the segregation of the chromosomes at meiotic division I [16]. The first eukaryotic RecA homologues, RAD51 and DMC1, were reported in budding yeast [17]. RAD51 expresses during mitosis and meiosis, whereas DMC1 expresses only during meiosis [18]. DMC1 disruption results in defects in reciprocal recombination, inability to form synaptonemal complexes, accumulation of DSBs, and abnormal chromosome synapses, showing D-loop formation and strand exchange activities [19]. The plant DMC1 orthologues were reported in *Lilium longiflorum* [20], *Arabidopsis thaliana* [21,22], *Oryza sativa* [23–26], and *Hordeum vulgare* [27]. The concept of deficient mutants in the meiotic recombination process arrests the meiotic recombination, crossing over, and pairing of chromosomes. This is the first step of reverse breeding to generate asynaptic mutants through which the parental-type gametes can be recovered in the progenies to fix the genomic composition and multiplication through TPS. Keeping this in mind, the present study aimed to silence the *StDMC1* gene in the Kufri Jyoti variety of potato to generate asynaptic mutants.

2. Materials and Methods

2.1. Plant Materials

The study was conducted using a popular tetraploid Indian potato variety, "Kufri Jyoti". Kufri Jyoti is an old potato variety (1968) developed by the ICAR-Central Potato Research Institute, Shimla, Himachal Pradesh, India, through clonal selection from a cross 3069d(4)/2814a(1). The variety has wide adaptability and is still popular among stakeholders. The virus-free in vitro tubes of Kufri Jyoti were obtained from the germplasm repository, Division of Crop Improvement and Seed Technology, ICAR-Central Potato Research Institute (CPRI), Shimla, Himachal Pradesh, India. The cultures were further multiplied in vitro on Murashige and Skoog media containing sucrose 30 g/L, calcium pantothenate 2 mL/L and gelrite 3 g/L). The pH of the media was adjusted to 5.7. A sufficient number of plants were generated for planting in a glass house. After 28 days of planting, leaf samples were collected for RNA isolation, and at the time of flowering, anthers and carpels were also collected for RNA isolation.

2.2. Identification of StDMC1 CDS, Sense and Anti-Sense Region

The *Arabidopsis thaliana AtDMC1* coding DNA sequence (CDS) was retrieved from the Arabidopsis Information Resource (TAIR) database. The potato *StDMC1* region was obtained by using the *AtDMC1* CDS as a query in the Potato Genome Sequencing Consortium (PGSC) database using the BLAST tool. The *StDMC1* CDS sequence was used to select the siRNA by using the bioinformatics tool siRNA-scan (http://bioinfo2.noble.org/RNAiScan.htm, accessed on 17 January 2021). Based on the CDS region, sense and antisense fragments were selected for RNAi construct development. The BLAST tool was used to ensure that there was no off-target silencing. The primers were designed using FastPCR version 4.0 [28] (Table 1).

Table 1. Details of primers used in this study. The red color denotes the restriction sites of different enzymes in the primer sequences.

Primer	Sequence with Restriction Enzyme	Size (bp)	Enzymes
StDMC1 Sense-F	CGCGTCGACGAAGATAGTGAACTTCGGC	526	*Sal* 1
StDMC1 Sense-R	CCCGGGTACCCTTTATCAATCGGGACAGC		*Kpn* 1
StDMC1 Antisense-F	CGCGGATCCGAAGATAGTGAACTTCGGC	433	*Bam* H1
StDMC1 Antisense-R	CCCGGGTACCAGAATCCACAATCAGAAGTC		*Kpn* 1
NPT II-F	TTTGTCAAGACCGACCTGTC	530	
NPT II-R	CCAACGCTATGTCCTGATAG		

2.3. Sense and Anti-Sense Amplification

The RNA isolation was performed by a NucleoSpin Macherey-Nagel RNA Plant kit (GmbH & Co. KG, Germany). The cDNA was synthesized using a High-Capacity cDNA kit by Applied Biosystems^{TM} (Waltham, MA, USA). The sense and anti-sense regions were amplified using the reaction: 1 µL cDNA, 15 µL EmeraldAmp PCR master mix (Takara^{TM}, Kusatsu, Japan), 1 µL of each forward and reverse primer, and 2 µL of RNase-free water to make a final reaction volume of 20 µL. The PCR conditions were as follows: 4 min at 95 °C followed by 35 cycles of 45 s at 94 °C, 45 s at 54 °C and 1 min at 72 °C, and terminated with one cycle of 5 min at 72 °C (Figure 1a). The same PCR conditions were followed for the antisense region except for the annealing temperature (52 °C). The amplified fragments were resolved on a 1.5% agarose gel (Figure 1b), and the eluted PCR products were sequenced using the BigDye v3.1 reaction kit, Applied Biosystems^{TM} (USA) and analyzed on the ABI 3500 Genetic Analyzer, Applied Biosystems^{TM} (USA).

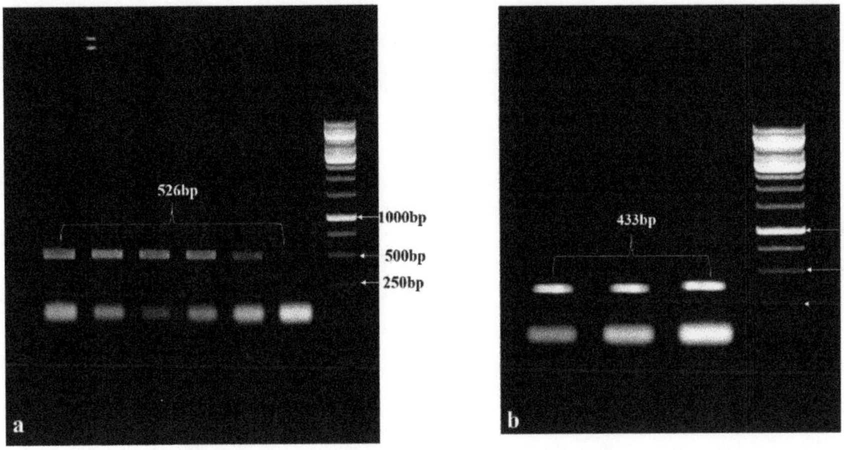

Figure 1. (**a**) PCR amplification of the *StDMC1* gene of a sense fragment with *StDMC* sense primers flanked with restriction sites *Sal* 1 and *Kpn* 1, resulting in a product size of 526 bp. (**b**) Amplification of an antisense gene fragment with *StDMC1* antisense primers flanked with restriction sites *Bam* HI and *Kpn* 1, resulting in a product size of 433 bp.

2.4. Development of siRNA Construct

The *StDMC1* sense fragment (526 bp) was amplified with gene-specific primers flanked with restriction sites for *Sal* 1 and *Kpn* 1, and the antisense gene fragment (433 bp) with restriction sites for *Bam* H1 and *Kpn* 1, respectively. These amplified PCR fragments were eluted from agarose gel and ligated in the pTZ57R/T TA cloning vector. The gene constructs pTZ57R/T::Sense *StDMC1* and pTZ57R/T::antisense were first cloned in *Escherichia coli* strain DH5α on a Luria Bertani (LB) agar plate containing Ampicillin (100 μg/mL). The colonies were confirmed via colony PCR as well as by restriction digestion. Later, these were subcloned into the binary vector pRI101AN. The sense plasmid and vector were digested with *Sal* 1 and *Kpn* 1 restriction enzymes for ligation. The ligated product was transformed using the heat shock method and maintained in *E. coli* strain DH5α selective media containing Kanamycin (50 μg/mL). The colonies obtained were confirmed through colony PCR using 35S forward and gene-specific reverse primers.

Similarly, the antisense fragment was amplified using antisense primers flanked with restriction sites for *Bam* H1 and *Kpn* 1 and subjected to restriction digestion for cloning in the *StDMC1* sense::pRI101AN vector backbone. The ligated products were maintained in *E. coli* strain DH5α. The confirmation of colonies obtained on selective LB agar plates containing Kanamycin was performed through colony PCR using 35S forward and NOST reverse primers. The whole cassette (*StDMC1*-sense::*StDMC1*-antisense::pRI101AN) was reconfirmed through restriction digestion using *Sal* 1 and *Bam* H1 enzymes.

2.5. Transformation of RNAi Construct in Agrobacterium Strain GV3101

The positive RNAi clone (*StDMC1*-sense::*StDMC1*-antisense::pRI101AN) was transformed into *Agrobacterium* strain GV3101 using the freeze-thaw method. *Agrobacterium* strain GV3101 carrying the RNAi gene cassette was grown in LB broth containing antibiotics (50 μg/mL of Kanamycin and 15 μg/mL of Rifampicin) in an incubator shaker at 28 °C for 48 h. From this culture broth, the plasmid was isolated using a Macherey-Nagel Nucleospin plasmid isolation kit (Germany). The transformed colonies were confirmed through restriction digestion using *Bam* H1 and *Hind* III enzymes.

2.6. *Agrobacterium*-Mediated Plant Transformation

The internodal stem cuttings of in vitro tissue-cultured plants were used as explant for genetic transformation. *Agrobacterium* culture harboring the *StDMC1* RNAi construct grown at 28 °C with an OD of 0.6, was used for transformation. For co-cultivation, the internodal cuttings were placed on pre-selective media containing basal MS medium for 48 h. Co-cultivation was performed with the *Agrobacterium* culture suspension obtained after centrifugation and resuspension in sterile MS liquid media. Acetosyringone (1 mg/mL) was added to the cell suspension containing cuttings to enhance virulence. The treated cuttings were dried using sterile tissue papers and then transferred to regeneration media containing Indoleacetic acid (IAA) (1 mg/mL), Naphthalene acetic acid (NAA) (0.01 mg/L), Gibberellic acid (GA3) (3 mg/mL), Cefotaxime (100 mg/mL), Kanamycin (50 µg/mL), and Carbenicillin (100 mg/L). The inoculated plates were kept at 20 ± 2 °C under fluorescent light at 100 µmol m^{-2} s^{-1} with 16 h of light and 8 h of dark until putative shoots emerged. These shoots were transferred into tubes containing MS media, Kanamycin (50 µg/mL), and IAA (300 µg/mL) for root initiation. The plantlets that survived were transplanted into the soil and grown under controlled conditions. Transformation efficiency was calculated using the following formula:

$$\text{Transformation efficiency } (\%) = \frac{\text{Number of positive transformants}}{\text{Total number of regenerated plantlets}} \times 100$$

2.7. *Screening of Putative Transformants*

DNA was isolated from the leaves of transformed plants using a QIAGEN DNeasy kit as per the manufacturer's instructions (North American manufacturing in Germantown, MD, USA). NPTII expression was tested by using NPTII forward and reverse primers. The PCR reaction contained 10 µL of Emerald PCR master mix (Takara™, Japan), 1 µL of each primer at a concentration of 1 µM, and 1 µL DNA templates in a reaction volume of 20 µL. PCR program was followed as 4 min at 94 °C; followed by 30 cycles 40 s at 94 °C, 50 s at 54 °C, 1 min at 72 °C, and 6 min at 72 °C. The transformed lines harboring the NPTII backbone were confirmed with an NPTII-specific primer on 1.3% agarose gel (Table 1).

2.8. *Pollen Viability Analysis*

The transformed lines as well as control plants were grown under controlled conditions. The flower buds were collected from transformed lines for pollen viability analysis. The pollen viability was tested using a 2% acetocarmine stain. An EVOS XL Core Imaging System (Invitrogen™, Waltham, MA, USA) microscope was used to examine the pollen viability. Five frames of each slide were examined to count the viable and non-viable pollen grains. Viable pollen grains retained the stain, whereas the non-viable pollen grains were unable to retain the stain. The percent viability was calculated by dividing the amount of stained pollen grains per field of view by the total number of pollen grains in that view.

2.9. *Quantitative Real-Time PCR (qRT-PCR)*

After the confirmation of pollen viability, non-viable lines along with Kufri Jyoti controls were used for quantitative real-time PCR (qRT-PCR). The qRT-PCR was performed in triplicates in reference to the housekeeping gene elf (elongation factor). The qRT-PCR reaction mixture contained 10 µL of SYBER GREEN® (Applied Biosystem™, USA), 1 µL cDNA, 1 µL of each forward and reverse primer, and the final volume was adjusted by adding 2 µL dH_2O (Table 2). Using the StepOnePlus™ Real-Time PCR System (Applied Biosystems™, USA), the expression of each gene, in comparison to the average Ct values of the housekeeping gene, was determined and analyzed. The $2^{-\Delta\Delta Ct}$ technique was used to quantitatively measure the relative changes in gene transcript levels [29].

Table 2. Details of primers used in the real-time PCR (qRT-PCR).

Real-Time Primer	Sequence
StDMC RT-F	TACATTACTGGGGAGTGAGGC
StDMC RT-R	CCCCAAAAGCTTCAGTTATTGC
elf-F	CGTTGTATCATGAATTTGTTTCTCTGT
elf-R	CCCCCTGAGGTTTCAACG

3. Results

3.1. Characterization of the DMC1 Gene in the Kufri Jyoti Potato Cultivar

The CDS sequence of the DMC1 gene from *Arabidopsis* was used as a query in the PGSC database (https://spuddb.uga.edu/index.shtml, accessed on 25 June 2021) via the BLAST tool. The CDS sequence of the *AtDMC1* gene showed 81% similarity with the diploid potato DMC1 CDS sequence. Primers designed from the gene sequence (Soltu.DM.09G025170.1) were used to amplify the DMC1 gene sequence in the tetraploid Kufri Jyoti cultivar. The sequence of Kufri Jyoti for the DMC1 gene was analyzed for sequence similarity using the NCBI nucleotide BLAST tool. The phylogenetic analysis carried out using MEGA 11.0.13 revealed that Kufri Jyoti's DMC1 has high sequence similarity with *Solanum lycopersicum* and *Capsicum* spp., followed by *Nicotiana tomentosiformis* and *N. tabacum* (Figure 2).

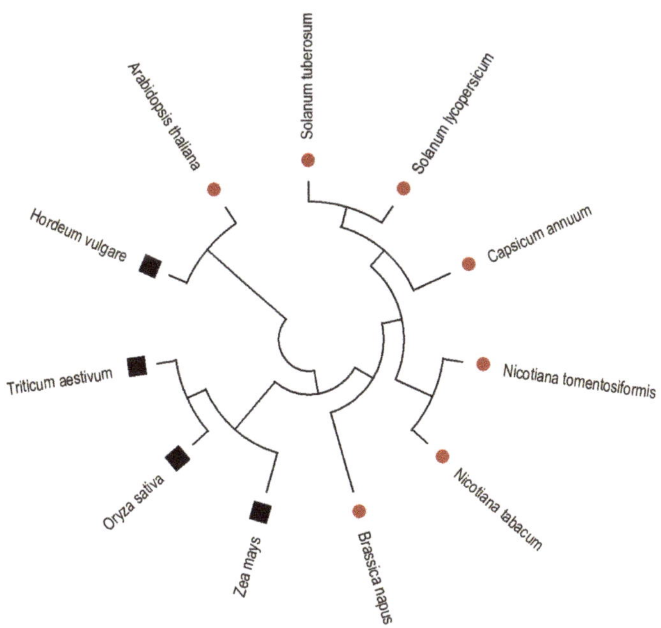

Figure 2. Phylogenetic analysis and comparison of the potato *StDMC1* gene sequence with that of major crops by the maximum likelihood method. In this phylogenetic tree, red color circles represent dicot species, and black squares represent monocot species.

3.2. Confirmation of pRI101 RNAi Construct

The sense fragment was confirmed through colony PCR. A band of 1.3 Kb gene size (526 bp *StDMC1* sense::801 bp::35S CaMV::58 bp 5′UTR) confirmed the insertion of the gene in the vector (Figure S1). Likewise, the antisense fragment was confirmed using colony PCR. A band of 1.8 Kb fragment size (801 bp 35S CaMV::526 bp Sense *StDMC1*::58 bp 5′UTR::433 bp *StDMC1* antisense::264 bp NOST) confirmed the successful integration of the full gene cassette in the pRI101AN vector (Figure S2). The whole cassette (*StDMC1* sense::*StDMC1* antisense::pRI101AN) was reconfirmed through restriction digestion using *Sal* 1 and *Bam* H1 (Figure 3). The sense fragment, being complementary and longer than the antisense fragment, would turn back and form a hairpin loop for knocking down the gene function, i.e., gene silencing.

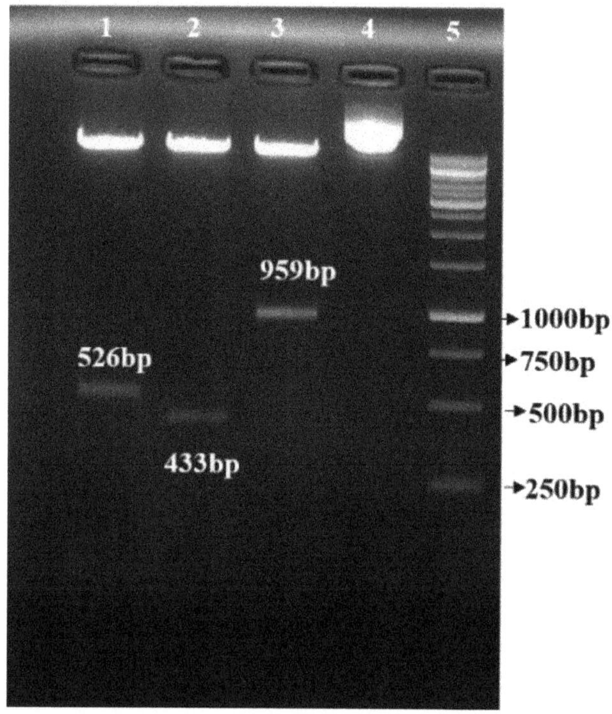

Figure 3. The plasmid containing the *StDMC1*:RNAi construct was digested with *Sal* 1 and *Kpn* 1 and *Bam* H1 and *Kpn* 1 restriction enzymes, respectively, to confirm sense and antisense insertion into the pRI101AN binary vector having the CaMV35S promoter and the NOS terminator. Lane 1: *StDMC1* sense with a product size of 526 bp digested with *Sal* 1 and *Kpn* 1, Lane 2: *StDMC1* antisense with a product size of 433 bp digested with *Bam* H1 and *Kpn* 1, Lane 3: Full cassette containing *StDMC1* sense::*StDMC1* antisense with a product size of 959 bp digested with *Sal* 1 and *Bam* H1, Lane 4: uncut pRI101AN plasmid with a product size of 10,417 bp, Lane 5: 1 Kb molecular marker.

3.3. *Agrobacterium-Mediated Plant Transformation*

Agrobacterium strain GV3101 harboring the complete RNAi cassette carrying *StDMC1* sense::*StDMC1* antisense::pRI101AN (Figure 4) was used to transform the Kufri Jyoti internodal explants for the generation of DMC1 mutated lines. The callus formation was observed 40 days after the co-cultivation, followed by putative shoot generation (Figure 5). In total, 310 putative transgenic lines were obtained, which were grown in a glass house under controlled conditions and were screened using the NPTII marker. Out of 310 lines, 137 lines were found positive, with the NPTII marker showing a product size of 530 bp (Figure 6). Overall, a transformation efficiency of 44% was obtained for mutants (Table 3).

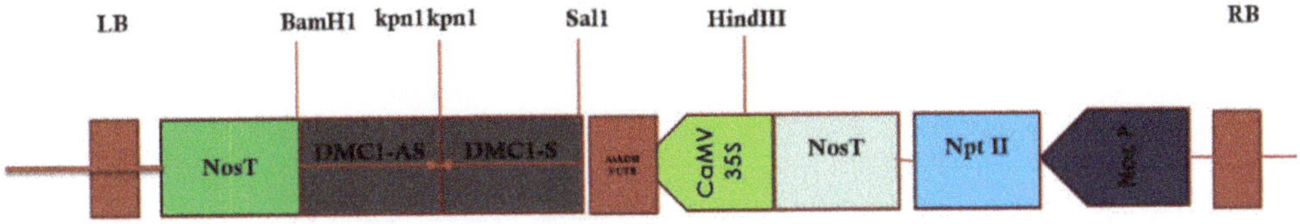

Figure 4. A schematic diagram of the RNAi vector construct. The hairpin structure consisting of a sense *StDMC1* sequence (oriented in *Sal* 1-*Kpn* 1 site) and the antisense *StDMC1* sequence (oriented in *Bam* HI-*Kpn* I site) was inserted between the (CaMV) 35S promoter and the nopaline synthase gene (NOS) terminator.

Figure 5. *Agrobacterium* harboring the *StDMC1* RNAi construct used for genetic transformation. (**a**) In vitro 15-day-old tissue culture internodal stem cuttings were placed on pre-selective basal MS medium for 48 h. (**b**) The treated cuttings were transferred to regeneration medium containing Indoleacetic acid (IAA) (1 mg/mL), Naphthalene acetic acid (NAA) (0.01 mg/L), Gibberellic acid (GA3) (3 mg/mL), Cefotaxime (100 mg/mL), Kanamycin (50 µg/mL), and Carbenicillin (100 mg/L) and plates were kept at 20 ± 2 °C under fluorescent light at 100 µmol m^{-2} s^{-1} with 16 h light and 8 h dark for callus formation. (**c**) Shoot development in putative transformants.

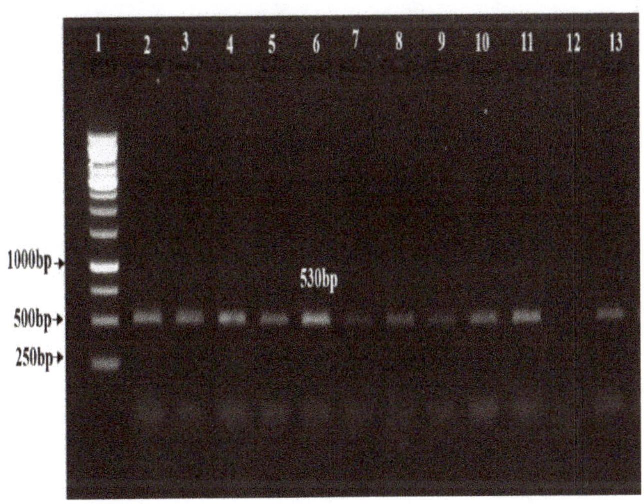

Figure 6. Validation of transformed lines by PCR amplification using NPTII primers resulting in a product size of 530 bp. Lane 1: 1 Kb Ladder, Lane 2–11: Transformed lines, Lane 12: Non-transformed line, i.e., Kufri Jyoti (control), and Lane 13: Positive plasmid containing *StDMC1* sense::*StDMC1* antisense:pRI101AN.

Table 3. Transformation efficiency of potato cultivar "Kufri Jyoti" genetically transformed with *Agrobacterium* strain GV3101 harboring binary vector pRI101.

Explant (Internode)	No. of Cuttings	Callus	No. of Regeneration of Shoots	NPTII Positive	Percentage of Transformation Efficiency
Kufri Jyoti	140	100	310	137	44

3.4. Pollen Viability

The acetocarmine-based pollen staining revealed that four DMC1 mutant lines had reduced pollen viability of 14–21% in comparison to the Kufri Jyoti control plants, which showed 78% pollen viability. Out of 137 positive lines, four mutant lines showed reduced pollen viability (Table 4; Figure 7). All other lines exhibited moderate-to-high pollen viability, as observed in the control Kufri Jyoti plants. The results confirmed the generation of DMC1 knock-out mutant lines.

Table 4. Pollen viability in putative DMC1 mutant lines.

Mutant Lines	Viable Pollens	Non-Viable Pollens	Total	% Viability
DMC4-37	18	107	125	14.4
DMC4-41	12	53	65	18.46
DMC6-20	26	97	123	21.14
DMC6-21	18	88	106	16.98
Kufri Jyoti- Control	112	31	143	78.32

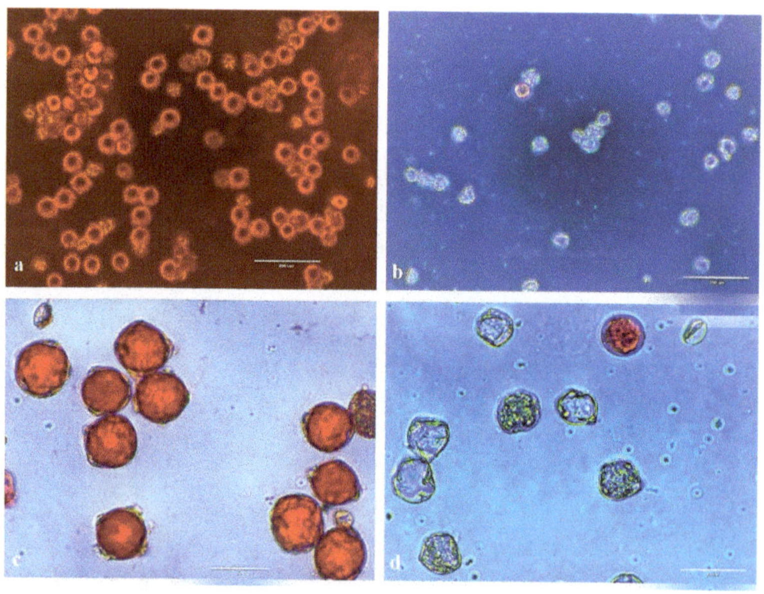

Figure 7. Pollen viability using acetocarmine (2%) under an EVOS XL Core Imaging System Microscope (Invitrogen^{TM}). Five frames of each slide were examined to count the viable and non-viable pollen grains. Viable pollen grains retained the stain, whereas the non-viable pollen grains were unable to retain the stain. (**a**) Control "Kufri Jyoti" pollen grains (scale bar = 200 µm), (**b**) RNAi lines showing non-viable pollen (scale bar = 200 µm), (**c**) control "Kufri Jyoti" pollen grains (scale bar = 50 µm), and (**d**) RNAi lines showing non-viable pollen (scale bar = 50 µm).

3.5. Quantitative Real-Time PCR (qRT-PCR)

qRT-PCR was carried out to measure the expression level of the DMC1 gene in mutant lines as well as in the control plant. The results of qRT-PCR clearly showed that DMC1 gene expression was significantly downregulated in four putative RNAi mutant lines in comparison to the control plant (Figure 8a). The expression of the DMC1 gene was downregulated to levels of 60-fold in DMC4-37, 56-fold in DMC4-41, 35-fold in DMC6-20, and 42-fold in DMC6-21. These findings support the pollen viability data and indicate the silencing of the DMC1 gene in these Kufri Jyoti lines.

Figure 8. Real-time quantitative reverse transcription polymerase chain reaction (qRT-PCR) analysis of control vs. RNAi lines. The expression of the *StDMC1* gene in comparison to the average Ct values of the housekeeping gene (elf) was determined and analyzed using the $2^{-\Delta\Delta Ct}$ method. (**a**) Expression analysis of control vs. putative mutant lines. (**b**) The fully grown control (left) and RNAi lines (right) plants of Kufri Jyoti.

4. Discussion

TPS could be a popular choice for the commercial production of potatoes, since it is simple to handle, transport, and is virus-free; however, due to its genome complexity and autotetraploid nature, it cannot remain true to type through sexual reproduction. Therefore, the production of non-recombinant gametes by arresting the crossing over among non-sister chromatids could be the first step towards achieving parental-type gametes for reconstitution of true-to-type varieties through sexual reproduction in potatoes. Here we aimed to develop non-recombinant gametes of the potato cultivar Kufri Jyoti by silencing the expression of the *StDMC1* gene through RNAi hairpin loop technology. At first, we identified the DMC1 gene in potato through in silico analysis of *Arabidopsis thaliana AtDMC1* CDS, which showed 81% sequence identity with the *StDMC1* CDS sequence (Soltu.DM.09G025170.1). We found only one DMC1 gene on chromosome 9 in the potato genome (http://spuddb.uga.edu/index.shtml, accessed on 25 June 2021), similar to barley [30], wheat [31], and maize [27]. Phylogenetic analyses revealed that the *StDMC1* gene of *S. tuberosum* had more similarity with *Solanaceous* crops, viz., *S. lycopersicum* and *Capsicum annum*, followed by *Nicotiana tomentosiformis* and *N. tabacum*; however, the DMC1 gene from monocot species formed a separate cluster except for *Hordeum vulgare*. The DMC1 nucleotide sequences of *S. tuberosum*, *S. lycopersicum*, and *C. annum*, which are all members of the same family, were found to be the most identical. The results provide strong evidence that the isolated cDNA homologue employed in this study is an accurate orthologue of the potato DMC1 gene.

The RNAi technique has been successfully used in potatoes to knock down the function of various genes, including the beta-carotene hydroxylase gene [20,32], susceptibility factors for late blight [33], and RNA viruses [34]. However, meiotic cycle genes have not been targeted in any autopolyploid to date. Although DMC1 and its orthologues have been studied for their functions in *Arabidopsis* [35], rice [23], wheat [36], and barley [37], this is the first attempt in any autopolyploid crop, the potato, to silence the function of the meiotic cycle gene *StDMC1*. We also chose the same strategy to knock down the function of the important meiotic cell division gene *StDMC1* in the popular Indian potato variety Kufri Jyoti as followed earlier in rice [25]. Kufri Jyoti is an old but widely adapted popular potato cultivar in India and has been successfully used in transformation studies previously [38]. The RNAi cassette consisting of sense and antisense strands (*StDMC1* sense::*StDMC1* antisense::pRI101AN) was successfully transformed into the potato variety Kufri Jyoti via *Agrobacterium*-mediated transformation, as has been demonstrated in an earlier study [39]. We used Kanamycin (50 mg/L) in the selection medium for shoot formation, while Deng et al. [25] used Hygromycin B (40 mg/L) in *Oryza sativa*. The transformation efficiency after screening with NPTII primers was found to be 44% in Kufri Jyoti putative mutants, which was considerably higher in comparison to 10–16% in Kufri Chipsona 1 cultivar [40].

Although there were no morphological differences observed between the mutant lines and control plants (Figure 8b), the gametic cells (pollen grains) showed reduced viability in mutants in comparison to control plants. Four mutant lines showed pollen viability of 14–21%, indicating partial sterility. It is well known that the plants with a defective DMC1 gene show reduced pollen viability compared to the control plants [25]. A similar type of study in *Arabidopsis* has already proven that silencing the DMC1 gene can alter the shape of pollens and reduce their viability [41]. Barley plants carrying mutations in the *HvDMC1* gene lead to unusual synapsis and chromosomal missegregation [25,42]. Likewise, Tian et al. [43] performed cytological studies to decipher the function of *AtDMC1* as well as ASY1 in *Arabidopsis* and showed that polyploidy can potentially alleviate lethal effects brought by mutations during plant reproductive life cycles. This indicates that DMC1 mutants may revert to normal meiosis in potato, which have an autopolyploid genome.

The analysis of putative four mutants through quantitative real-time PCR (qRT-PCR) revealed substantial down-regulation of DMC1 gene expression in four mutant lines, DMC4-37, DMC4-41, DMC6-20, and DMC6-21, in comparison to control Kufri Jyoti plants, indicating that the RNAi construct was successful in silencing the DMC1 gene in these transformed lines. Similar down-regulation and low transcript levels for the DMC1 gene have been observed earlier in maize calli transformed by pART-DmRNAi [27].

The results revealed and verified the role of *StDMC1* in potatoes, which acts as a platform for the generation of non-recombinant gametes and could play an important role in the development of a TPS-based breeding program in potatoes. However, these lines need further characterization in advanced sexual generations for use as asynaptic mutants in potatoes. We have also summarized our results in Figure 9, which illustrates the experimental procedure for the silencing of the *StDMC1* gene in potatoes.

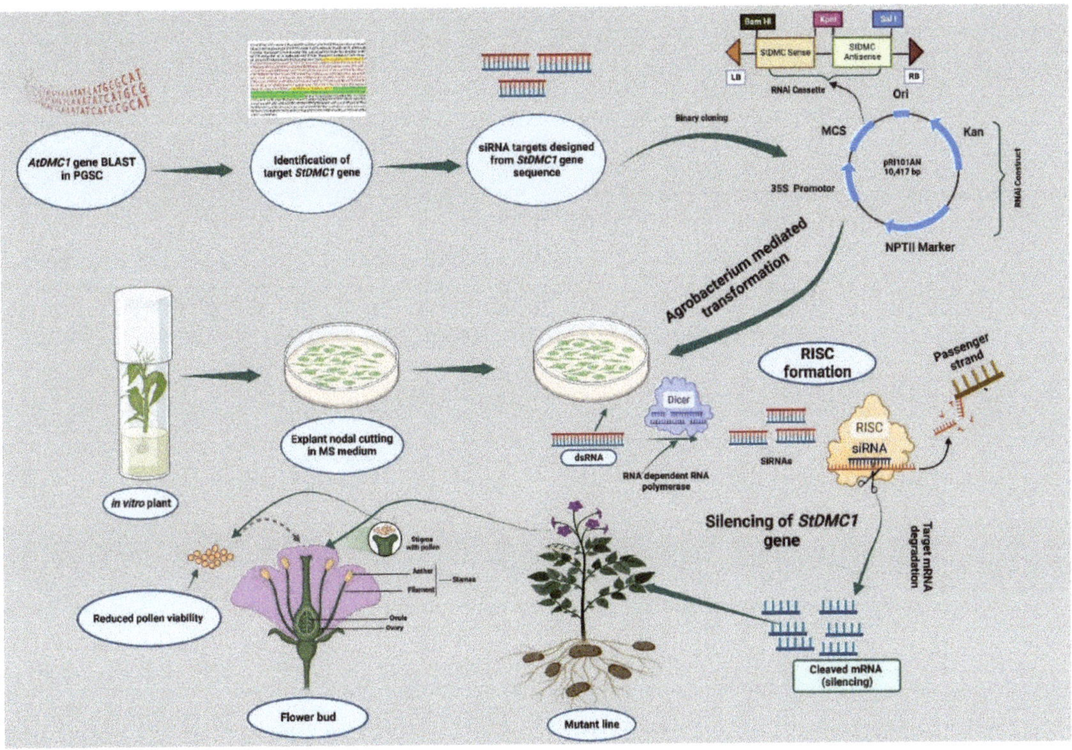

Figure 9. Schematic illustration of the mechanism of silencing of *StDMC1* gene through RNA interference (RNAi) technique. *Arabidopsis thaliana AtDMC1* CDS sequence was used as a query in the Potato Genome Sequencing Consortium (PGSC) database and the *StDMC1* gene sequence was retrieved for designing the siRNA target through in silico analysis. The target gene sequence was cloned in a binary vector, i.e., pRI101AN, and transferred in plants through *Agrobacterium*-mediated transformation. The process included shearing dsRNA into smaller dsRNA intermediates by the RNase-III type endonuclease known as Dicer. The small interfering RNA (siRNA) containing the complementary sequence of the target RNA acts as a guide RNA, and the other strand, known as the passenger RNA, gets degraded. The mutant lines were obtained by using internodal explant cutting on Kanamycin-selective MS medium. Reduced pollen viability was observed in silenced *StDMC1* gene lines.

5. Conclusions

To prevent cross-over and recombination in potatoes, we used the RNAi method to target the production of *StDMC1* mutants. The CaMV35S promoter was employed to direct the creation of an RNAi construct using an efficient siRNA. The pollen viability of putative mutants was lower than that of the control, Kufri Jyoti plants. Four mutant lines showing partial pollen sterility also exhibited low expression levels of *StDMC1* in comparison to the control. These four lines will be further investigated in sexual generations for use as asynaptic mutants in potatoes. This is the first study on knocking down the function of the *StDMC1* gene in auto-tetraploid potato to develop asynaptic mutants. The characterization of four mutant lines generated in this study could open new research opportunities to study asynaptic mutants of potatoes. Our findings show that meiosis in *S. tuberosum* can be controlled to produce meiotic mutants and parental lines for TPS breeding. We also show how RNAi-mediated meiosis modification can pave the way for the creation of effective plant breeding techniques in potatoes.

Supplementary Materials: The following supporting information can be downloaded at: https://www.mdpi.com/article/10.3390/life13010174/s1. (Figure S1: Validation of *StDMC1*:sense fragment in pRI101AN binary vector using Colony PCR. The transformants were confirmed with 35S forward and gene specific reverse primer. The 1.3 Kb gene fragment (526 bp *StDMC1*:Sense + 801 bp 35S CaMV + 58 bp 5′UTR) confirmed the insertion of gene in vector and Figure S2: Validation of *StDMC1* sense::*StDMC1* antisense fragment in pRI101AN binary vector using Colony PCR. The transformants were confirmed with 35S forward and NOST reverse primers. The 1.8 Kb gene fragment (526 bp *StDMC1*: sense + 433 bp *StDMC1* antisense + 801 bp 35S CaMV + 58 bp 5′UTR) confirmed the insertion of gene in vector).

Author Contributions: Conceptualization, S.S. (Salej Sood) and U.G.; Investigation, A.K.; Methodology, A.K. and S.S. (Sundaresha Siddappa); Resources, S.S. (Salej Sood), V.B. and V.K.; Supervision, S.S. (Salej Sood) and U.G.; Writing—original draft, A.K.; Writing –assistance, S.S. (Sundaresha Siddappa), D., B.S., N.S. and B.D. Writing—review & editing, S.S. (Salej Sood). All authors have read and agreed to the published version of the manuscript.

Funding: This research received no external funding.

Institutional Review Board Statement: Not applicable.

Informed Consent Statement: Not applicable to the current research as we have not used any humans/or animal subjects.

Data Availability Statement: The author confirms that the data supporting the finding of this study are available within the article and its Supplementary Materials.

Acknowledgments: We are thankful to Kameshwar Sen, Shruti Gupta, Nimisha Kaushal, Rahul and Vandana Kashyap for help in plant tissue culture facility. We are also thankful to Director ICAR-CPRI for facilitation during the study.

Conflicts of Interest: The authors declare no conflict of interest.

References

1. Guchi, E. Disease Management Practice on Potato (*Solanum tuberosum* L.) in Ethiopia. *World J. Agric. Res.* **2015**, *3*, 34–42. [CrossRef]
2. Singh, B.; Goutam, U.; Kukreja, S.; Siddappa, S.; Sood, S.; Sharma, J.; Bhardwaj, V. Biofortification Strategies to Improve Iron Concentrations in Potato Tubers: Lessons and Future Opportunities. *Potato Res.* **2021**, *65*, 51–64. [CrossRef]
3. Bhaskar, P.B.; Venkateshwaran, M.; Wu, L.; Ané, J.M.; Jiang, J. Agrobacterium-Mediated Transient Gene Expression and Silencing: A Rapid Tool for Functional Gene Assay in Potato. *PLoS ONE* **2009**, *4*, e5812. [CrossRef]
4. Vreugdenhil, D.; Bradshaw, J.; Gebhardt, C.; Govers, F.; Mackerron, D.; Taylor, M.A.; Ross, H.A. *Potato Biology and Biotechnology*; Elsevier: Amsterdam, The Netherlands, 2007; pp. 157–177, ISBN 9780444510181. [CrossRef]
5. Bonierbale, M.W.; Amoros, W.R.; Salas, E.; de Jong, W. Potato Breeding. In *The Potato Crop Its Agricultural, Nutritional and Social Contribution to Humankind*; Springer: Berlin/Heidelberg, Germany, 2019; pp. 163–217. [CrossRef]
6. Vos, P.G.; Uitdewilligen, J.G.A.M.L.; Voorrips, R.E.; Visser, R.G.F.; van Eck, H.J. Development and Analysis of a 20K SNP Array for Potato (*Solanum tuberosum*): An Insight into the Breeding History. *Theor. Appl. Genet.* **2015**, *128*, 2387–2401. [CrossRef] [PubMed]
7. Perrino, E.V.; Perrino, P. Crop wild relatives: Know how past and present to improve future research, conservation and utilization strategies, especially in Italy: A review. *Genet. Resour. Crop Evol.* **2020**, *67*, 1067–1105. [CrossRef]
8. Sood, S.; Mangal, V.; Bhardwaj, V.; Kardile, H.; Sharma, A.K. Diploid F1 Hybrid TPS Potato Breeding Pipeline and Prospects. *Potato J.* **2021**, *48*, 89–92.
9. Bradshaw, J.E. Improving the Nutritional Value of Potatoes by Conventional Breeding and Genetic Modification. In *Quality Breeding in Field Crops*; Springer: Berlin/Heidelberg, Germany, 2019; pp. 41–84. [CrossRef]
10. Kumari, M.; Kumar, M.; Solankey, S.S. Breeding Potato for Quality Improvement. In *Potato—From Incas to All Over World*; IntechOpen: London, UK, 2018. [CrossRef]
11. Kandemir, N.; Saygili, İ. Apomixis: New Horizons in Plant Breeding. *Turk. J. Agric. For.* **2015**, *39*, 549–556. [CrossRef]
12. Sood, S.; Kumar, A.; Sundaresha, S.; Bhardwaj, V. Genome Editing Prospects to Develop Disease/Pest-Resistant Potato Varieties. In *Sustainable Management of Potato Pests Diseases*; Springer: Berlin/Heidelberg, Germany, 2022; pp. 413–434. [CrossRef]
13. Brown, M.S.; Bishop, D.K. DNA Strand Exchange and RecA Homologs in Meiosis. *Cold Spring Harb. Perspect. Biol.* **2015**, *7*, a016659. [CrossRef] [PubMed]
14. Symington, L.S.; Rothstein, R.; Lisby, M. Mechanisms and Regulation of Mitotic Recombination in Saccharomyces Cerevisiae. *Genetics* **2014**, *198*, 795–835. [CrossRef] [PubMed]
15. Crickard, J.B.; Greene, E.C. The Biochemistry of Early Meiotic Recombination Intermediates. *Cell Cycle* **2018**, *17*, 2520–2530. [CrossRef]

16. Shinoharaa, A.; Shinohara, M. Roles of RecA Homologues Rad51 and Dmc1 during Meiotic Recombination. *Cytogenet. Genome Res.* **2004**, *107*, 201–207. [CrossRef]
17. Lin, Z.; Kong, H.; Nei, M.; Ma, H. Origins and Evolution of the RecA/RAD51 Gene Family: Evidence for Ancient Gene Duplication and Endosymbiotic Gene Transfer. *Proc. Natl. Acad. Sci. USA* **2006**, *103*, 10328–10333. [CrossRef] [PubMed]
18. Kurzbauer, M.T.; Uanschou, C.; Chen, D.; Schlögelhofer, P. The Recombinases DMC1 and RAD51 Are Functionally and Spatially Separated during Meiosis in Arabidopsis. *Plant Cell* **2012**, *24*, 2058–2070. [CrossRef]
19. Reitz, D.; Grubb, J.; Bishop, D.K. A Mutant Form of Dmc1 That Bypasses the Requirement for Accessory Protein Mei5-Sae3 Reveals Independent Activities of Mei5-Sae3 and Rad51 in Dmc1 Filament Stability. *PLoS Genet.* **2019**, *15*, e1008217. [CrossRef] [PubMed]
20. Kobayashi, T.; Kobayashi, E.; Sato, S.; Hotta, Y.; Miyajima, N.; Tanaka, A.; Tabata, S. Characterization of CDNAs Induced in Meiotic Prophase in Lily Microsporocytes. *DNA Res.* **1994**, *1*, 15–26. [CrossRef]
21. Sato, S.; Hotta, Y.; Research, S.T.-D. Structural Analysis of a RecA-like Gene in the Genome of *Arabidopsis thaliana*. *DNA Res.* **1995**, *2*, 89–93. [CrossRef] [PubMed]
22. Doutriaux, M.P.; Couteau, F.; Bergounioux, C.; White, C. Isolation and Characterisation of the RAD51 and DMC1 Homologs from *Arabidopsis thaliana*. *Mol. Gen. Genet. MGG* **1998**, *257*, 283–291. [CrossRef] [PubMed]
23. Ding, Z.J.; Wang, T.; Chong, K.; Bai, S. Isolation and Characterization of OsDMC1, the Rice Homologue of the Yeast DMC1 Gene Essential for Meiosis. *Sex. Plant Reprod.* **2001**, *13*, 285–288. [CrossRef]
24. Metkar, S.S.; Sainis, J.K.; Mahajan, S.K. Cloning and Characterization of the DMC1 Genes in *Oryza sativa*. *Curr. Sci.* **2004**, *87*, 353–357.
25. Deng, Z.Y.; Wang, T. OsDMC1 Is Required for Homologous Pairing in *Oryza sativa*. *Plant Mol. Biol.* **2007**, *65*, 31–42. [CrossRef]
26. Sakane, I.; Kamataki, C.; Takizawa, Y.; Nakashima, M.; Toki, S.; Ichikawa, H.; Ikawa, S.; Shibata, T.; Kurumizaka, H. Filament Formation and Robust Strand Exchange Activities of the Rice DMC1A and DMC1B Proteins. *Nucleic Acids Res.* **2008**, *36*, 4266–4276. [CrossRef]
27. Etedali, F.; Kohnehrouz, B.B.; Valizadeh, M.; Gholizadeh, A. Genome Wide Cloning of Maize Meiotic Recombinase Dmc1 and Its Functional Structure through Molecular Phylogeny. *Genet. Mol. Res.* **2011**, *10*, 1636–1649. [CrossRef]
28. Kalendar, R.; Lee, D.; Schulman, A.H. FastPCR Software for PCR, in Silico PCR, and Oligonucleotide Assembly and Analysis. *Methods Mol. Biol.* **2014**, *1116*, 271–302. [CrossRef]
29. Livak, K.J.; Schmittgen, T.D. Analysis of Relative Gene Expression Data Using Real-Time Quantitative PCR and the 2(-Delta Delta C(T)) Method. *Methods* **2001**, *25*, 402–408. [CrossRef]
30. Barakate, A.; Higgins, J.D.; Vivera, S.; Stephens, J.; Perry, R.M.; Ramsay, L.; Colas, I.; Oakey, H.; Waugh, R.; Franklin, F.C.H.; et al. The Synaptonemal Complex Protein ZYP1 Is Required for Imposition of Meiotic Crossovers in Barley. *Plant Cell* **2014**, *26*, 729–740. [CrossRef] [PubMed]
31. Devisetty, U.K.; Mayes, K.; Mayes, S. The RAD51 and DMC1 Homoeologous Genes of Bread Wheat: Cloning, Molecular Characterization and Expression Analysis. *BMC Res. Notes* **2010**, *3*, 245. [CrossRef] [PubMed]
32. Sundaresha, S.; Jeevalatha, A.; Kumar, R.; Sood, S.; Sharma, S.; Bhardwaj, V.; Singh, B.; Chakrabarti, S.K. RNA Interference: A Versatile Tool to Augment Plant Protection Strategies in Potato. In *Sustainable Management of Potato Pests and Diseases*; Springer: Berlin/Heidelberg, Germany, 2022; pp. 369–388. [CrossRef]
33. Sun, K.; Wolters, A.M.A.; Vossen, J.H.; Rouwet, M.E.; Loonen, A.E.H.M.; Jacobsen, E.; Visser, R.G.F.; Bai, Y. Silencing of Six Susceptibility Genes Results in Potato Late Blight Resistance. *Transgen. Res.* **2016**, *25*, 731–742. [CrossRef] [PubMed]
34. Hameed, A.; Tahir, M.N.; Asad, S.; Bilal, R.; van Eck, J.; Jander, G.; Mansoor, S. RNAi-Mediated Simultaneous Resistance Against Three RNA Viruses in Potato. *Mol. Biotechnol.* **2017**, *59*, 73–83. [CrossRef]
35. Da Ines, O.; Degroote, F.; Goubely, C.; Amiard, S.; Gallego, M.E.; White, C.I. Meiotic Recombination in Arabidopsis Is Catalysed by DMC1, with RAD51 Playing a Supporting Role. *PLoS Genet.* **2013**, *9*, e1003787. [CrossRef]
36. Draeger, T.; Martin, A.C.; Alabdullah, A.K.; Pendle, A.; Rey, M.D.; Shaw, P.; Moore, G. Dmc1 Is a Candidate for Temperature Tolerance during Wheat Meiosis. *Theor. Appl. Genet.* **2020**, *133*, 809–828. [CrossRef]
37. Szurman-Zubrzycka, M.; Baran, B.; Stolarek-Januszkiewicz, M.; Kwaśniewska, J.; Szarejko, I.; Gruszka, D. The Dmc1 Mutant Allows an Insight into the DNA Double-Strand Break Repair during Meiosis in Barley (*Hordeum vulgare L.*). *Front. Plant Sci.* **2019**, *10*, 761. [CrossRef] [PubMed]
38. Sanju, S.; Siddappa, S.; Thakur, A.; Shukla, P.K.; Srivastava, N.; Pattanayak, D.; Sharma, S.; Singh, B.P. Host-Mediated Gene Silencing of a Single Effector Gene from the Potato Pathogen *Phytophthora Infestans* Imparts Partial Resistance to Late Blight Disease. *Funct. Integr. Genomics* **2015**, *15*, 697–706. [CrossRef] [PubMed]
39. Tomar, G.; Chakrabarti, S.K.; Sharma, N.N.; Jeevalatha, A.; Sundaresha, S.; Vyas, K.; Azmi, W. RNAi-Based Transgene Conferred Extreme Resistance to the Geminivirus Causing Apical Leaf Curl Disease in Potato. *Plant Biotechnol. Rep.* **2018**, *12*, 195–205. [CrossRef]
40. Kaur, A.; Guleria, S.; Reddy, M.S.; Kumar, A. A Robust Genetic Transformation Protocol to Obtain Transgenic Shoots of *Solanum tuberosum* L. Cultivar 'Kufri Chipsona 1'. *Physiol. Mol. Biol. Plants* **2020**, *26*, 367–377. [CrossRef] [PubMed]

41. Wijnker, E.; Deurhof, L.; van de Belt, J.; de Snoo, C.B.; Blankestijn, H.; Becker, F.; Ravi, M.; Chan, S.W.L.; van Dun, K.; Lelivelt, C.L.C.; et al. Hybrid Recreation by Reverse Breeding in *Arabidopsis thaliana*. *Nat. Protoc.* **2014**, *9*, 761–772. [CrossRef] [PubMed]
42. Colas, I.; Barakate, A.; MacAulay, M.; Schreiber, M.; Stephens, J.; Vivera, S.; Halpin, C.; Waugh, R.; Ramsay, L. Desynaptic5 Carries a Spontaneous Semi-Dominant Mutation Affecting Disrupted Meiotic CDNA 1 in Barley. *J. Exp. Bot.* **2019**, *70*, 2683–2698. [CrossRef] [PubMed]
43. Tian, B.M.; Yan, B.; Gao, J.Y.; Si, Y.H.; Zang, X. Dissecting 2 Meiotic Mutations (Dmc1 and Asy1) in Artificial Allopolyploid *Arabidopsis thaliana*. *Cytologia* **2011**, *76*, 411–423. [CrossRef]

life

MDPI

Article

Analysis of Rac/Rop Small GTPase Family Expression in *Santalum album* L. and Their Potential Roles in Drought Stress and Hormone Treatments

Yu Chen [1,†], Shengkun Wang [2,†], Xiaojing Liu [1], Dongli Wang [2], Yunshan Liu [1], Lipan Hu [1,*] and Sen Meng [2,*]

1 College of Biology and Food Engineering, Chongqing Three Gorges University, Chongqing 404100, China
2 State Key Laboratory of Tree Genetics and Breeding, Research Institute of Tropical Forestry, Chinese Academy of Forestry, Guangzhou 510520, China
* Correspondence: xinonghulipan@163.com (L.H.); mengsen021124@126.com (S.M.)
† These authors have contributed equally.

Abstract: Plant-specific Rac/Rop small GTPases, also known as Rop, belong to the Rho subfamily. Rac proteins can be divided into two types according to their C-terminal motifs: Type I Rac proteins have a typical CaaL motif at the C-terminal, whereas type II Rac proteins lack this motif but retain a cysteine-containing element for membrane anchoring. The *Rac* gene family participates in diverse signal transduction events, cytoskeleton morphogenesis, reactive oxygen species (ROS) production and hormone responses in plants as molecular switches. *S. album* is a popular semiparasitic plant that absorbs nutrients from the host plant through the haustoria to meet its own growth and development needs. Because the whole plant has a high use value, due to the high production value of its perfume oils, it is known as the "tree of gold". Based on the full-length transcriptome data of *S. album*, nine *Rac* gene members were named *SaRac1-9*, and we analyzed their physicochemical properties. Evolutionary analysis showed that SaRac1-7, AtRac1-6, AtRac9 and AtRac11 and OsRac5, OsRacB and OsRacD belong to the typical plant type I Rac/Rop protein, while SaRac8-9, AtRac7, AtRac8, AtRac10 and OsRac1-4 belong to the type II Rac/ROP protein. Tissue-specific expression analysis showed that nine genes were expressed in roots, stems, leaves and haustoria, and *SaRac7/8/9* expression in stems, haustoria and roots was significantly higher than that in leaves. The expression levels of *SaRac1*, *SaRac4* and *SaRac6* in stems were very low, and the expression levels of *SaRac2* and *SaRac5* in roots and *SaRac2/3/7* in haustoria were very high, which indicated that these genes were closely related to the formation of *S. album* haustoria. To further analyze the function of *SaRac*, nine *Rac* genes in sandalwood were subjected to drought stress and hormone treatments. These results establish a preliminary foundation for the regulation of growth and development in *S. album* by *SaRac*.

Keywords: *Santalum album*; *Rac* gene family; haustorium; tissue-specific expression; drought stress; hormone treatments

Citation: Chen, Y.; Wang, S.; Liu, X.; Wang, D.; Liu, Y.; Hu, L.; Meng, S. Analysis of Rac/Rop Small GTPase Family Expression in *Santalum album* L. and Their Potential Roles in Drought Stress and Hormone Treatments. *Life* **2022**, *12*, 1980. https://doi.org/10.3390/life12121980

Academic Editor: Francois Lefort

Received: 20 October 2022
Accepted: 24 November 2022
Published: 26 November 2022

1. Introduction

S. album is a small semiparasitic tree belonging to the genus *Santalum* in the family *Santalaceae*. This family is composed of 29 genera with approximately 400 species, 19 of which are specific to the *Santalum* genus [1–3]. It is distributed in India, Malaysia, Australia and Indonesia and is also cultivated in Guangdong and Taiwan [4]. *S. album* is known as the "tree of gold", and its heartwood is widely used in high-end craft sculpture and high-end furniture manufacturing [5,6]. The essential oil extracted from its heartwood is mainly used in the cosmetic and pharmaceutical industries due to its special aroma [7,8]. Sandalwood usually yields 3–7% essential oil depending on the region and hemisphere [9]. The value of a sandalwood tree depends on three important characteristics: the volume of heartwood and the concentration and quality of its heartwood oil [10]. *S. album* grows for a long time; generally, it begins to form heartwood after 7 or 8 years of growth in natural environments. Unsustainable exploitation of wild trees, combined with increasing global demand for sandalwood products has threatened native sandalwood populations in some places, such as southern India. Because of the contradiction between the increasing market demands for its heartwood and the shortage of *S. album* heartwood on the market, the shortage of heartwood with its destruction of natural resources has become increasingly complicated [11].

S. album is a semiparasitic plant; its roots have a special organ, the haustorium, which, through the haustorium's contact with the host plant root, absorbs its own water and nutrients [12]. Therefore, the growth and development of haustoria play a vital role in the growth and development of *S. album* and the formation of heartwood. As far as the current scientific progress is concerned, the study of the regulatory genes related to haustoria in *S. album* is still limited.

ROS can regulate plant growth and development, hormone transduction, root hair development and so on. In sandalwood, Rboh is significantly induced by the haustorial inducer DMBQ, and the ROS signal produced by Rboh protein is necessary for the development of sandalwood haustoria [13]. At present, the growth and development of sandalwood is related to the size, quantity and quality of haustoria. According to the available literature, Rac and Rboh can regulate ROS concentrations through interactions and then promote the formation of sandalwood haustoria. In conclusion, the interaction between Rac and Rboh is an indispensable link in the formation of ROS for haustorium development.

Plant-specific Rac small GTPases belong to the plant Rho subfamily [14]. In animals, Rho is divided into several subfamilies, including Rho, Cdc42 and Rac. Rop is also called Rop in plants [15]. Rac is the sole plant subfamily of the conserved Rho family of small GTPases [16–19]. They are soluble proteins that localize and function at the plasma membrane by way of posttranslational lipid modifications [20–22]. Rac proteins can be divided into two types according to their C-terminal motifs: Type I Rac proteins, which have a typical CaaL motif at the C-terminal, and Type II Rac proteins, which have a truncated and functional motif. Whereas type-I Racs probably undergo prenylation, type-II Racs undergo S-acylation but not prenylation [23]. All type-II Racs have an additional exon at the 3′ end of the gene, probably resulting from the insertion of an intron into an ancestral Rac [24,25]. Most if not all plant Racs act at the plasma membrane [26], and they are molecular switches that regulate diverse signaling cascades [27,28]. They are widely involved in plant signal transduction [29–31], cytoskeleton morphogenesis, polarized cell growth [32,33], cell morphogenesis, defense, responses and reactive oxygen species (ROS) production [34–36]. Taken together, Rac small GTPases act as a simple binary switch capable of receiving a wide variety of inputs and accordingly generating a multitude of specific outputs. Some members of the plant *Rac* gene family have been shown to regulate the growth and development of *S. album* through ROS production. Based on previous studies, nine *Rac* genes were obtained from the transcriptome and genomic data of *S. album* and compared with eleven *Rac* family members in *A. thaliana* and seven *Rac* family members in rice. The bioinformatics data of these sequences and tissue-specific expression were analyzed and these results will provide a basis for further analysis of the related functions of *S. album Rac* genes and its relationship with haustorium formation. Additionally, the expression levels of *SaRac1-9* under drought stress and hormone treatments were studied, which play an important role in the growth and development of sandalwood.

2. Materials and Methods

2.1. Plant Materials and Treatments

The materials used in the experiment are stored in this laboratory, Institute of Tropical Forestry, Chinese Academy of Forestry Sciences (Guangzhou, 23°11′ N, 113°23′ E). Four tissues were collected: root, stem, leaf and primary haustorium (after collection, they were immediately immersed in liquid nitrogen), each with three biological replicates, and stored in a −80 °C refrigerator. Sandalwood seedlings grew in a greenhouse in the soil for approximately 2 months, until they reached 15 cm. Saplings displaying similar growth were subsequently used for analysis. The soil water was withheld to drought for 0, 3 and 9 d, in which the soil relative water content was reduced from 70% to 32%. The soil relative water content of control plants was maintained at 80%, which represents a suitable soil environment for sandalwood growth. For hormone treatments (ABA, IAA, ethephon), the plants were divided into four zones (ABA, IAA, ethephon, CK), which were treated with different hormones at a concentration of 100μM/L. Three biological replicates were included. After 48 h, samples were collected at the same time and immediately immersed in liquid nitrogen.

2.2. Genome-Wide Identification and Sequence Analysis of Rac Genes in S. album

According to the gene annotation information of the sandalwood transcriptome, the coding sequence of *Rac* was obtained, and the DNA sequence and promoter sequence of *Rac* were obtained from *S. album* genome data by homologous alignment. The nucleotide sequence homology of *Rac* in sandalwood was analyzed using TBtools.

The conserved domains of nine SaRac proteins were predicted using NCBI-Conserved Domain Search, and the structure of SaRac was analyzed using TBtools. Novel putative motifs were explored using the MEME server (https://meme-suite.org/meme/, accessed on 13 July 2022). By selecting a motif of full length, eight conserved motifs located in the SaRac domain were identified. To analyze the cis-regulatory element in the putative *SaRac* promoter, we performed a cis-regulatory element analysis of the 2000 BP promoter sequence of the nine *SaRac* genes using TBtools and the Plant CARE website (http://bioinformatics.psb.ugent.be/webtools/plantcare/html/, accessed on 1 August 2022), and their functions and numbers were predicted.

2.3. Analysis of Physicochemical Properties of SaRac

The basic physical and chemical properties of the protein, including molecular weight (MW), isoelectric point (pI) hydrophilicity and number of amino acids, were analyzed using ProtParam online software (https://web.expasy.org/protparam/, accessed on 25 August 2022). Protein subcellular localization was predicted by using the WoLF PSORT website (https://wolfpsort.hgc.jp/, accessed on 27 August 2022).

2.4. Phylogenetic Analysis

A multiple sequence alignment of 27 Rac proteins from *S. album* and other species, including *A. thaliana* and rice, was performed. The 11 Rac members of *A. thaliana* were obtained from TAIR (https://www.arabidopsis.org/, accessed on 12 July 2022), and the 7 Rac members of rice were obtained from NCBI (https://www.ncbi.nlm.nih.gov/, accessed on 12 July 2022) using the MUSCLE method. A phylogenetic tree was constructed by using the NJ method implemented in MEGA-X. The parameters for tree construction were as follows. Phylogenetic analysis parameters: bootstrap (1000 replicates); gaps/missing data: pairwise deletion; model: Dayhoff model; pattern among lineages: same (homogeneous) and rates among sites: uniform rates. Finally, the phylogenetic tree was constructed.

2.5. Chromosome Location and Gene Structure Analysis

Positional information and gene structures of *SaRac* genes on chromosomes of *S. album* were obtained from the gene annotation information of the *S. album* transcriptome. The chromosomal locations were displayed with TBtools (github.com/CJ-Chen/TBtools, accessed on 30 August 2022). The numbers and organization of introns, exons and gene structures were drawn and displayed using TBtools.

2.6. Collinearity of the SaRac Gene

The chromosomal locations of *SaRac* genes were obtained by TBtools software. Using TBtools software, the synteny relationships of orthologous *Rac* genes among *S. album*, *A. thaliana* and rice were evaluated. The parameters for collinearity of the *SaRac* gene were as follows: the genome file was used to construct the chromosome skeleton to obtain the gene density file, and then TBtools was used to prepare the collinearity file, extract the location of the *SaRac* gene, highlight the collinearity region of the target gene and finally, visualize it.

2.7. Expression Profiles of SaRac Genes in Different Tissues and under Drought and Hormone Treatments

According to the instructions of the plant RNA extraction kit (R6827, Omega Bio-tek, Norcross, GA, USA), total RNA was extracted from *S. album*. The *SaRac1-9* gene sequence was obtained by reverse transcription of cDNA with a reverse transcription kit (DRR037S, Takara, Dalian, China), and quantitative primers were designed according to the whole-field sequence of the *SaRac1-9* gene (Table S1). Real-time quantitative PCR primers were designed using Primer 3.0 software with *S. album*. Actin was used as an internal reference gene. Using cDNA as a template, real-time quantitative PCR was performed according to the instructions of the SYBR Green Premix Ex Taq II Kit (Qiagen, Dusseldorf, Germany). The qPCR mixture were 0.5 μL of primers, 1 μL of cDNA, 10 μL of 2 × Mix and 8 μL of dd H_2O. The reaction procedure was 95 °C for 10 s, 60 °C for 10 s and 72 °C for 20 s, and 40 cycles were performed. Three biological replicates and 3 technical replicates were carried out. SPSS 25 software was used to analyze the significance of the data, and OriginPro 2019b software was used to draw the graph.

3. Results

3.1. Sequence Analysis of the SaRac Gene Family in S. album

Nine cDNA and DNA sequences of *SaRac* were isolated from the transcriptome and genome of *S. album* and named *SaRac1-9*. The basic physical and chemical properties of the protein were analyzed by ProtParam (Table 1). The nine predicted full-length Rac proteins varied from 196 (SaRac5) to 211 (SaRac7) amino acid residues, and the relative molecular mass ranged from 21.45 (SaRac5) to 23.28 (SaRac7) kDa, with isoelectric points in the range of 9.18–9.62.

Table 1. Chemical properties of proteins in the *SaRac* gene family.

Gene Name	Name	Number of Amino Acids	Molecular Weight (kDa)	PI	Instability Index	Total Number of Negatively Charged Residues (Asp + Glu)	Total Number of Positively Charged Residues (Arg + Lys)	Grand Average of Hydro-pathicity (GRAVY)	In Silico Prediction WOLF PSORT
Sal10G07090.1	SaRac1	197	21.58	9.20	38.58	18	25	−0.075	plas
Sal10G04200.1	SaRac2	198	21.83	9.43	34.09	16	27	−0.129	chlo
Sal8G02490.1	SaRac3	198	21.81	9.38	40.23	18	25	−0.042	chlo
Sal7G05910.1	SaRac4	197	21.55	9.32	41.48	18	27	−0.106	chlo
Sal9G31620.1	SaRac5	196	21.45	9.25	39.09	16	28	−0.047	chlo
Sal9G09920.1	SaRac6	196	21.74	9.62	38.96	17	25	−0.083	chlo
Sal6G04230.1	SaRac7	211	23.28	9.18	44.44	19	28	−0.069	chlo
Sal9G07020.1	SaRac8	209	22.94	9.27	36.73	19	27	−0.141	plas
Sal9G04490.1	SaRac9	196	21.63	9.55	40.28	18	27	−0.120	cyto

3.2. Analysis of SaRac Protein Homology

The conserved domains of the nine SaRac proteins were predicted using NCBI-Conserved Domain Search, and the results showed that all nine Racs contained Rop-like domains (Figure 1A) belonging to the *Rop* gene family. The structure of the *SaRac* gene family was analyzed using TBtools (Figure 1B). To investigate the structural diversity of *Racs*, a total of 10 conserved motifs in the *Racs* were captured by the MEME website, and we obtained five conserved motifs located in the Rac domain. We further analyzed the sequence structures of nine *SaRac* genes and aligned them to eleven *Rac* members in *A. thaliana* [37] and seven *Rac* members in rice [38] (Figure 1C,D). It is worth noting that the type and distribution of the C-terminal domains of most *Racs* are similar. The results showed that the nine SaRac proteins all had five conserved motifs, which were the same as the Rac proteins in *Oryza sativa* and *A. thaliana*.

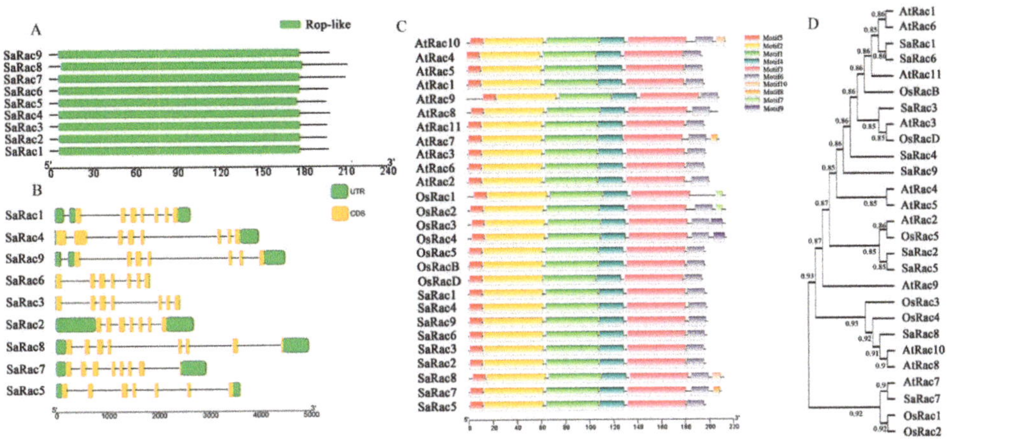

Figure 1. The gene structures and conserved motifs of Rac family members in *S. album* based on evolutionary relationships. (**A**) Rop-like domains of SaRac proteins. (**B**) The exon-intron structure of SaRac proteins. (**C**) Conserved motifs of SaRac proteins. (**D**) Tree of evolutionary relatives: rice, *A. thaliana* and sandalwood.

3.3. Sequence Analysis of the SaRac Gene Family

Novel putative motifs were explored using the MEME server. By selecting a motif of full length, we identified eight conserved motifs located in the *SaRac* domain (Figure 2A,B).

To analyze cis-acting elements in putative *SaRac* promoters, the 2000 BP promoter sequences of nine *SaRac* genes were identified as cis-regulatory elements by TBtools and the Plant CARE website, and their functions and numbers were predicted (Figure 2C). There were differences in the types and numbers of regulatory elements in the promoters of the nine *SaRac* genes, but all of them had multiple hormone response elements and stress response elements. For example, *SaRac1* and *SaRac4* have two hormone response elements, while *SaRac2* and *SaRac3* have three hormone response elements, *SaRac6-8* have four hormone response elements and *SaRac5* and *SaRac9* have five hormone response elements. In addition, *SaRac3/6/8* have one stress response element, *SaRac4/7/9* have two stress response elements and *SaRac5* and *SaRac2* have three and four stress response elements, respectively. Among them, *SaRac6* has ten CGTCA motifs, and we can further speculate that this gene may be related to the chemical defense response of *S. album*.

3.4. Chromosome Distribution of the SaRac Gene Family

To clarify the distribution of *SaRac* on the chromosomes of *S. album*, we used TBtools software to map the location of *SaRac* family members (Figure 3A). The *SaRac* family members of *S. album* showed irregular distribution on the chromosome and did not form a large number of gene clusters. The graph shows that nine *SaRac* genes are located on chr6, chr7, chr8, chr9 and chr10 of *S. album*, and more than half of them have one *SaRac* gene member on the chromosome. The remaining two chromosomes have four and one *SaRac* gene members. Notably, in *S. album* linkage groups6/7/8/9/10, the eight *SaRac* genes were classified into five segmental duplication events (*SaRac4*/*SaRac1*, *SaRac4*/*SaRac3*, *SaRac3*/*SaRac1*, *SaRac9*/*SaRac1* and *SaRac6*/*SaRac1*) (Figure 3A,B) (Table S2).

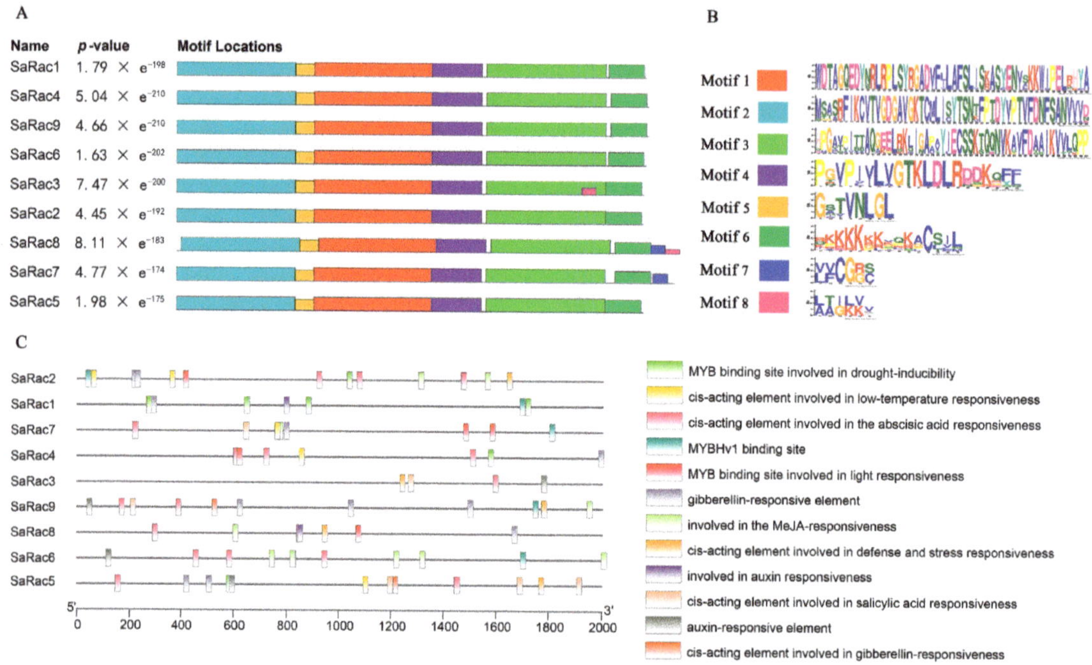

Figure 2. Sequence analysis of the *SaRac* gene family. (**A**) Motif locations of SaRac. (**B**) Discovered motifs. (**C**) Cis-acting Elements in Putative *SaRac* Promoters.

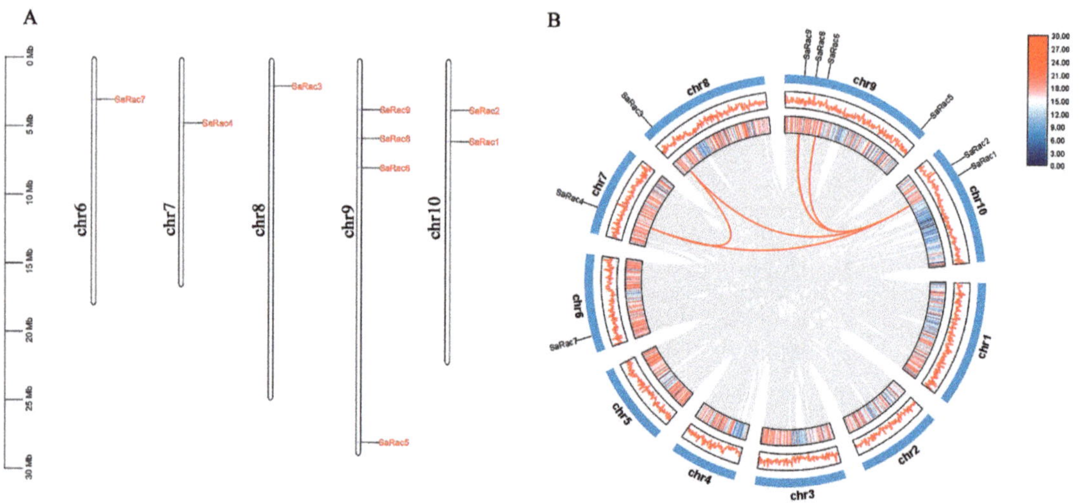

Figure 3. Interchromosomal relationships of *SaRac* genes. (**A**) The location of *SaRac* genes on the *S. album* chromosome. (**B**) Interchromosomal relationships of *Rac* genes in *S. album*. Gray lines indicate all synteny blocks in the *S. album* genome, and red lines indicate segmental duplication events.

To further infer the phylogenetic relationship between sandalwood, *A. thaliana* and rice, we constructed two syntenic maps of sandalwood with *A. thaliana* and rice. A total of three *SaRac* genes showed syntenic relationships with *Racs* in *O. sativa* (Figure 4A). In *A. thaliana*, there are many *SaRac* genes that are in common with *Racs* (Figure 4B) (Table S3).

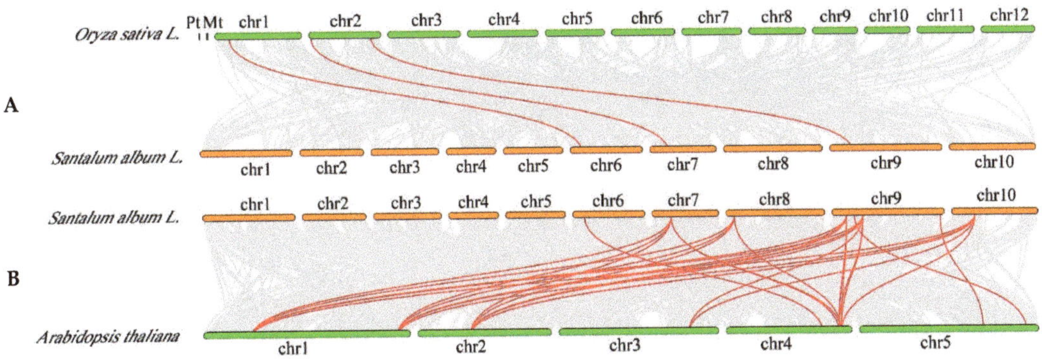

Figure 4. Synteny analysis of *Rac* genes between sandalwood and two representative plant species. (**A**,**B**) Gray lines in the background indicate the collinear blocks within sandalwood and other plant genomes, while the red lines highlight the syntenic *Rac* gene pairs.

3.5. Phylogenetic Comparison of Rac in Different Species

To further evaluate the phylogenetic relationship between SaRac and other plants, 11 and 7 Rac sequences from *A. thaliana* and rice were compared with 9 Rac sequences from *S. album*, and the phylogenetic tree of the whole protein sequence alignment was constructed by MEGA-X. The results showed that the *Rac* genes of these three species can be divided into five subgroups: I, II, III, IV and V (Figure 5). From the available literature, we further obtained the role of *Rac* in *A. thaliana* and rice. For instance, in *A. thaliana*, *AtRac1/6/11* are highly expressed in mature pollen and play an important role in pollen tube growth [37]. *AtRac2* overexpression inhibits the growth of root tips. *AtRac3/8* inhibit ABA-induced responses, including actin recombination in guard cells, stomatal closure, seed germination, root elongation and gene expression [39,40]. *AtRac4* is a positive regulator of root hair initiation and apical growth [41]. *AtRac5* acts on actin dynamics, polar growth, root hair growth and so on; finally, *AtRac10* participates in the regulation of membrane transport. On the other hand, in rice, *OsRac1* is a resistance and grain size gene [42,43]. *OsRac4/5* are negative regulators of blast resistance, and OsRacB is a direct effector of OsRopGEF2/3/6. It is a potential downstream target of OsRopGEF2/3/6/8 and confers salt tolerance, a negative regulator of disease resistance [44]. In addition, we can more accurately estimate the functions likely to be contained in the members of the nine *SaRac* gene families.

3.6. Analysis of Rac/Rop Multibase Regions in S. album, A. thaliana and O. sativa

The members of the *Rac* gene family are divided into type I Rac/Rop protein and type II Rac/Rop protein. Type I proteins have a conserved CaaL motif at the C-terminus. However, type II has a truncated, but functionally modified, posttranscriptional motif. Nine SaRac, eleven AtRac and seven OsRac proteins were sequenced, and the results showed that SaRac1-7 and AtRac1-6, AtRac9 and AtRac11 as well as OsRac5, OsRacB and OsRacD belong to the typical type I, while SaRac8-9 and AtRac7, AtRac8 and AtRac10 and OsRac1-4 belong to type II (Figure 6).

The conserved G domains in the N-terminal region of these proteins were GTPase active domains (G1, G3), Mg^{2+} binding domains (G2) and GTP binding sites (G4, G5). The G2 and G3 domains are also called switch I and II loops, and the C21 and C156 positions are the conserved L-cysteine residues of the G domain (Figure 6).

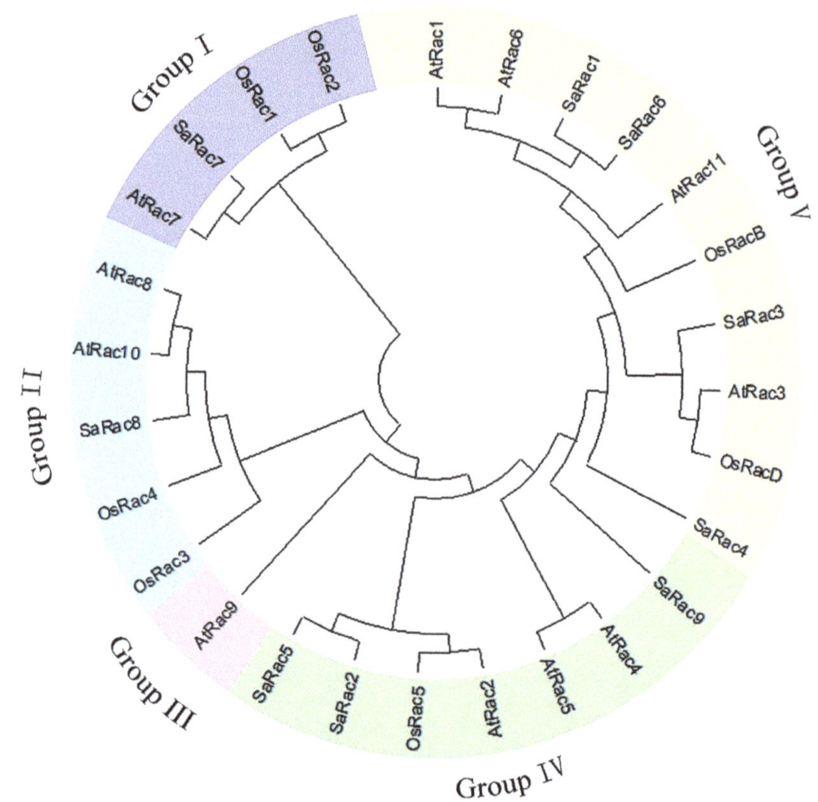

Figure 5. Phylogenetic analysis of Racs from three species (*A. thaliana*, *O. sativa*, *S. album*). Full-length polypeptide sequences were used to make the interspecific phylogenetic tree.

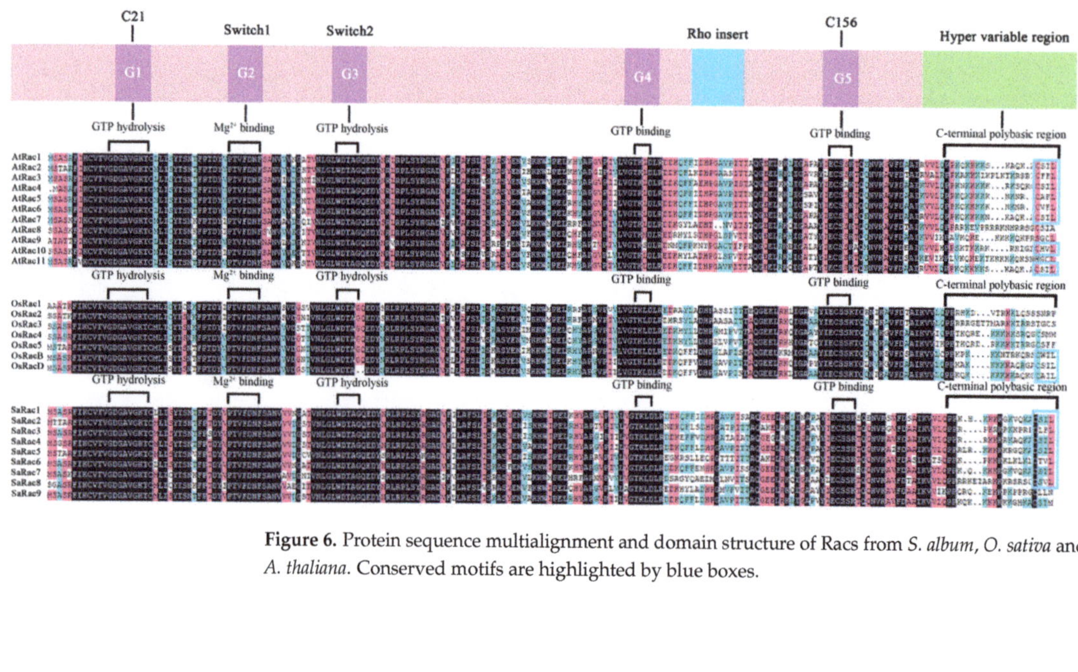

Figure 6. Protein sequence multialignment and domain structure of Racs from *S. album*, *O. sativa* and *A. thaliana*. Conserved motifs are highlighted by blue boxes.

3.7. Tissue Specificity of Rac Expression in S. album

We first investigated the tissue-specific expression of nine *Rac* gene family members in roots, stems, leaves and primary haustoria of *S. album* by green fluorescent quantitative PCR with gene-specific primers. Nine genes (*SaRac1-9*) were expressed in roots, stems, leaves and primary haustoria, but there was a difference in their expression levels. The expression of *SaRac7/8/9* in stems, haustoria and roots was higher than that in leaves, but the expression of *SaRac1*, *SaRac4* and *SaRac6* in stems was lower in leaves. The expression levels of *SaRac2* in stems and primary haustoria were high. It is worth noting that the expression levels of *SaRac2* and *SaRac5* in roots were approximately 6 and 13, respectively, relative to leaves (Figure 7). Therefore, we can further speculate that *SaRac2* and *SaRac5* may have a strong positive correlation with the growth and development of haustoria in sandalwood. In contrast, a high expression of *SaRac2* and *SaRac5* may inhibit the expression of *SaRac2* and *SaRac5* in leaves. The tissue-specific expression of *Rac* showed that different *Rac* members play different roles in different signaling pathways of *S. album*.

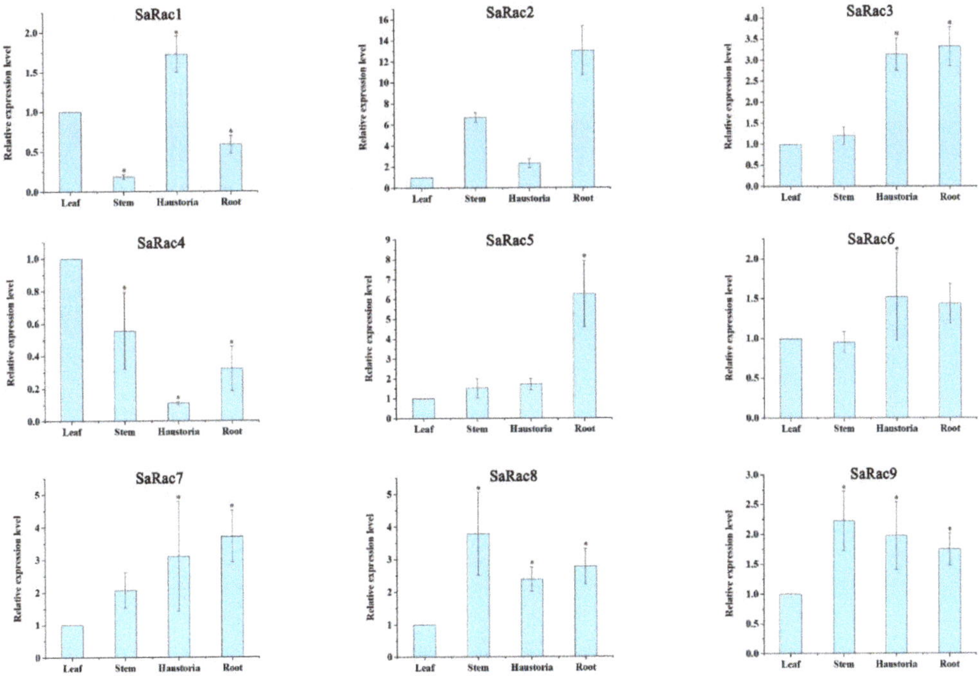

Figure 7. Expression profiles of *SaRac* genes in *S. album* across different organs. The expression level of *SaRac* genes in *S. album* in four organs (leaf, stem, haustoria, root). The relative expression level was calculated by setting the expression value of *SaRac* genes in the leaves of *S. album* at 1. * indicates significant difference by *t* test at $p < 0.05$.

3.8. Expression of SaRac Genes under Drought and Hormones Treatments

To better understand the function of *SaRac* in response to abiotic stress, nine *SaRac* genes were selected for further analysis. Under drought stress, the expression of more than a third of the genes, including *SaRac1*, *SaRac3*, *SaRac4* and *SaRac7* were increased. The general trend of *SaRac1*, *SaRac3*, *SaRac4* and *SaRac7* expression was first increased and finally decreased at 9 d. However, the expression levels of *SaRac2* and *SaRac9* were downregulated. Only two genes, *SaRac6* and *SaRac8*, showed higher expression under long-term drought treatment than those under control conditions (Figure 8A).

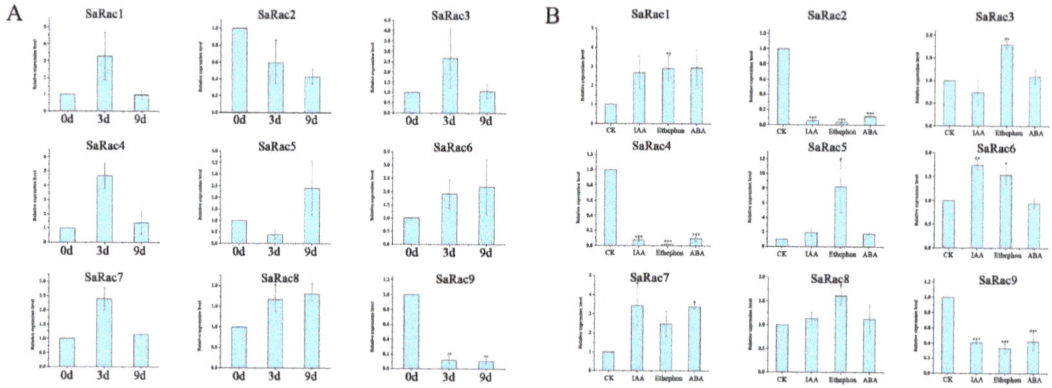

Figure 8. Expression of *SaRac* genes under drought and hormone treatments. (**A**) Expression of *SaRac* genes under drought stress. (**B**) Expression of *SaRac* genes in response to hormones. * indicates significant difference by t test at $p < 0.05$; ** indicates significant difference by t test at $p < 0.01$; *** indicates significant difference by t test at $p < 0.001$.

The expression levels of *SaRac1, SaRac6, SaRac7* and *SaRac8* increased after 48 h of IAA treatments, indicating that these genes were responsive to IAA. On the contrary, the expression levels of *SaRac2, SaRac4, SaRac5* and *SaRac9* were lower under IAA treatments than those under control conditions. Furthermore, more than half of the genes, including *SaRac1, SaRac3, SaRac5* and *SaRac6,* showed higher expression under ethephon treatments. It is worth mentioning that the expression of *SaRac5* under ethephon-treated conditions was 8 times higher than that under control conditions. Moreover, ABA treatment significantly induced the expression of *SaRac1, SaRac3, SaRac7* and *SaRac8*. However, *SaRac2, SaRac4* and *SaRac9* expression were decreased in response to both ethephon and ABA treatments.

4. Discussion

Rac small GTPases are members of the plant-specific Rho subfamily and are involved in many signaling events, such as defense responses, pollen tube growth, root hair development, reactive oxygen species (ROS) production and phytohormone response, and play a very important role in the abovementioned events [45,46]. Rac protein is a soluble protein that localizes in the plasma membrane and functions through posttranslational lipid modification [23,47,48]. For example, 11 *Rac* genes have been identified in *A. thaliana,* and there are eight type I *Rac* genes: *AtRac1-6, AtRac9,* and *AtRac11*. Seven *Rac* gene family members were identified in rice, while seven *Rac* family members were identified in *H. vulgare,* indicating that the copy number of *Rac* genes is not the same in different species. Studies have shown that *AtRac1/6/11* are highly expressed in mature pollen, and *AtRac4/2* are a pair of positive and negative regulators of root hair tips. At present, most of the research on *Rac* genes focuses on medicine and animals, while research on plants focuses on model plants, such as *A. thaliana, O. sativa* and *Hordeum vulgare.* Little is known about the mechanisms by which *SaRac* impacts the growth of *S. album.*

Reactive oxygen species (ROS) produced by NADPH oxidase have been shown to play many important roles in signaling and development in plants, such as in plant defense response, cell death, abiotic stress, stomatal closure, and root hair development [49–53]. In sandalwood, they control the development and formation of haustoria [13]. At present, studies have shown that the interaction between Rac GTPases and the N-terminal extension is ubiquitous and that a substantial part of the N-terminal region of Rboh, including the two EF-hand motifs, is required for the interaction [54]. The interaction between Rac and Rboh also provides further theoretical help for the study of the mechanism of Rac. At the same time, the regulation of Rboh ROS production by Rac provides a theoretical basis for the development of sandalwood haustoria.

Auxin, abscisic acid (ABA) and ethephon play key roles in the development of many plants. These three hormones are often used as the main substances in plant hormone response experiments. In this study, we treated nine *Rac* genes with drought and hormones. Only two genes, *SaRac6* and *SaRac8*, showed higher expression under long-term drought stress than those under control conditions.

The results of hormone treatments indicated that the expression of *SaRac1*, *SaRac3*, *SaRac7* and *SaRac8* were higher under ABA treatments. Moreover, more than half of the genes, including *SaRac1*, *SaRac3*, *SaRac5* and *SaRac6*, showed higher expression under ethephon treatments. Furthermore, the expression levels of *SaRac1*, *SaRac6*, *SaRac7* and *SaRac8* increased after 48 h of IAA treatments, indicating these genes are responsive to IAA. Previous studies had demonstrated that auxin biosynthesis is essential for haustorium in haustorium formation in the root-parasitic plants [55,56]. Thus, these genes may be involved in the ontogeny of the *S. album* haustorium and further influence the growth and development of sandalwood.

In our study, to better understand the evolution of the *Rac* gene family in sandalwood, the structure, conserved motifs, phylogenetic relationships and collinearity of *SaRac* genes were characterized. One conserved motif was located in the Rac domain, suggesting that the Rac domain is conserved among *A. thaliana*, rice and sandalwood. Most of the *SaRac* genes exhibited similar numbers of exons. Phylogenetic analysis revealed that SaRac, AtRac and OsRac proteins could be classified into five subgroups. Groups IV and V contain more Racs that come from all three species, and groups I and II also contain Racs from all three species. Moreover, group III contains one Rac in *A. thaliana*. For instance, SaRac5 to SaRac9 in group IV and SaRac4 to AtRac1 in group V could have expansive functions surrounded by AtRac, SaRac or OsRac proteins with different functions.

Additionally, synteny maps between two representative species and sandalwood were constructed to better understand the phylogenetic relationships. More than 10 pairs were detected in *A. thaliana*, and three pairs were detected in rice, indicating a strong homologous relationship between sandalwood and *A. thaliana*, and a weak homologous relationship between sandalwood and rice.

Gene duplication is a major mechanism underlying the evolution of novel protein functions. We detected five *SaRac* genes that were assigned to segmental duplication events, implying high segmental duplication. These results indicated that some *SaRac* genes were possibly generated by gene duplication.

Most Rac proteins identified to date have been functionally characterized in *A. thaliana* and rice, and their roles include the regulation of root hair initiation and apical growth, hormonal responses, stress responses and so on. Among them, the best-described Rac proteins are the members of group IV (Figure 5). These *Rac* genes are involved in root hair formation and disease resistance. In the phylogenetic tree, we can more accurately estimate the functions likely to be contained in the members of the *SaRac* gene family. For instance, we can predict the role of *SaRac* gene by the *AtRac* members of group IV. Similarly, we can make a preliminary prediction of the role of *SaRac* members in groups I/II/V.

Sandalwood is considered one of the most valuable trees in the world. Its heartwood is often used in carving crafts, cosmetics, medicine and other industries, but its main value lies in the essential oils extracted from the heartwood [57]. Therefore, it is important to investigate whether *SaRac* may be related to their accumulation in sandalwood haustorium tissue, thereby affecting the growth and development of the heartwood of sandalwood and, in turn, the quality of heartwood essential oils [58].

In gene expression, promoters play an important role in regulation, through which gene expression can be changed to change the characteristics of plants. Therefore, the study of promoters is a key step in genetic engineering and gene expression research [59]. In addition to the typical core promoter, there are many regulatory elements controlling the functional expression of genes. Analysis of cis-acting elements revealed that most of the *SaRac* promoters contained a large number of elements related to hormones and stress response. The abscisic acid response element (ABRE) was found in the promoters of

SaRac2-9, and the gibberellin response element (P-box) was found in the promoters of *SaRac1/2/4/5/6/7/8/9* However, *SaRac2/3/5/8/9* all had TC-rich repeats related to defense and stress responses, suggesting that these *Rac* genes may be involved in disease resistance and stress resistance of *S. album*.

To understand the function of nine *SaRac* members more accurately, based on the functional prediction of gene promoters, in this study, the expression levels of nine genes at four sites were detected by green fluorescent quantitative PCR. The results showed that, compared with the expression levels of leaves in each gene, the expression levels of *SaRac1*, *SaRac4* and *SaRac6* in stems were very low, while the expression levels of *SaRac2* in stems and primary haustoria were very high, and the expression levels of *SaRac2* and *SaRac5* in roots were approximately 6 and 13, respectively. In general, the level of protein expression in specific tissues is closely related to its function. The high expression of *SaRac2* and *SaRac5* in roots and *SaRac2/3/7* in haustoria may indicate that these genes are closely related to the formation of haustoria in *S. album*.

5. Conclusions

Because of the lack of research on the molecular mechanism of growth and development in *S. album* at present, in this study, the biological information and expression pattern of 9 *Rac* genes in *S. album* were analyzed (Figure 9). These results lay the physical and chemical foundation for further studies of the *Rac* family genes involved in the growth and development of *S. album* and regulation its functions. The perspectives of research on the semiparasitic sandalwood will develop towards the Rac-dependent generation of ROS in promoting haustorium development, which more effectively provides a data base for the growth mechanism of sandalwood.

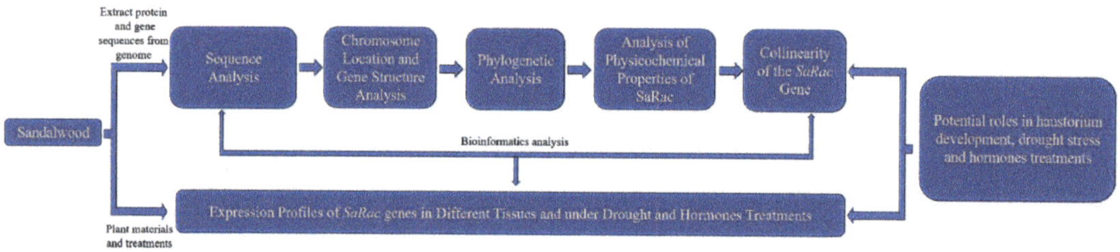

Figure 9. Framework figure. Protein and gene sequences of *SaRacs* were obtained from genome. Bioinformatic analyses were conducted, and the expression profiles of *SaRac* genes in different tissues and under drought and hormones treatments were obtained. These results established a preliminary foundation for the functional study of *SaRac* genes.

Supplementary Materials: The following supporting information can be downloaded at: https://www.mdpi.com/article/10.3390/life12121980/s1, Table S1: Primer design; Table S2: Segmental replication events in sandalwood; Table S3: Synteny analysis of Rac genes between sandalwood, A. thaliana and rice.

Author Contributions: Conceptualization, Y.C. and S.M.; methodology, Y.C. and X.L.; validation, D.W. and Y.L.; formal analysis, Y.C. and S.M.; investigation, Y.C. and S.W.; resources, S.M. and L.H.; data curation, S.M.; writing—original draft preparation, Y.C.; writing—review and editing, Y.C. and X.L.; funding acquisition, S.M. and L.H. All authors have read and agreed to the published version of the manuscript.

Funding: This work was supported by grants from the Natural Science Foundation of Chongqing, China (CSTC2018JCYJAX0778), the Science and Technology Research Program of Chongqing Municipal Education Commission (KJQN202101245), the National Natural Science Foundation of China (31722012 and 3190130), Natural Science Foundation of Guangdong Province, China (2019A1515011595) and Fundamental Research Funds for the Central Non-profit Research Institution of CAF (CAFYBB 2019QD001 and CAFYBB2020SY018).

Institutional Review Board Statement: Not applicable.

Informed Consent Statement: Not applicable.

Data Availability Statement: Not applicable.

Conflicts of Interest: The authors declare no conflict of interest.

References

1. Harbaugh, D.T. A taxonomic revision of Australian northern sandalwood (*Santalum lanceolatum, Santalaceae*). *Aust. Syst. Bot.* **2007**, *20*, 409–416. [CrossRef]
2. Harbaugh, D.T.; Oppenheimer, H.L.; Wood, K.R.; Wagner, W.L. Taxonomic revision of the endangered Hawaiian red-flowered sandalwoods (*Santalum*) and discovery of an ancient hybrid species. *Syst. Bot.* **2010**, *35*, 827–838. [CrossRef]
3. Butaud, J.F. Reinstatement of the Loyalty Islands Sandalwood, Santalum austrocaledonicum var. glabrum (*Santalaceae*), in New Caledonia. *PhytoKeys* **2015**, *56*, 111–126. [CrossRef]
4. Harbaugh, D.T.; Baldwin, B.G. Phylogeny and biogeography of the sandalwoods (*Santalum, Santalaceae*): Repeated dispersals throughout the Pacific. *Am. J. Bot.* **2007**, *94*, 1028–1040. [CrossRef] [PubMed]
5. Jones, C.G.; Ghisalberti, E.L.; Plummer, J.A.; Barbour, E.L. Quantitative co-occurrence of sesquiterpenes; a tool for elucidating their biosynthesis in Indian sandalwood, *Santalum album*. *Phytochemistry* **2006**, *67*, 2463–2468. [CrossRef]
6. Viollon, C.; Chaumont, J.P. Antifungal properties of essential oils and their main components upon Cryptococcus neoformans. *Mycopathologia* **1994**, *128*, 151–153. [CrossRef]
7. Benencia, F.; Courrèges, M.C. Antiviral activity of sandalwood oil against herpes simplex viruses-1 and -2. *Phytomedicine* **1999**, *6*, 119–123. [CrossRef] [PubMed]
8. Teixeira da Silva, J.A.; Kher, M.M.; Soner, D.; Page, T.; Zhang, X.; Nataraj, M.; Ma, G. Sandalwood: Basic biology, tissue culture, and genetic transformation. *Planta* **2016**, *243*, 847–887. [CrossRef]
9. Demole, E.; Demole, C.; Enggist, P. A chemical investigation of the volatile constituents of east Indian sandalwood oil (*Santalum album* L.). *Helv. Chim. Acta.* **1976**, *59*, 737–747. [CrossRef]
10. Subasinghe, U.; Gamage, M.; Hettiarachchi, D.S. Essential oil content and composition of Indian sandalwood (*Santalum album*) in Sri Lanka. *J. For. Res.* **2014**, *24*, 127–130. [CrossRef]
11. Diaz-Chavez, M.L.; Moniodis, J.; Madilao, L.L.; Jancsik, S.; Keeling, C.I.; Barbour, E.L.; Ghisalberti, E.L.; Plummer, J.A.; Jones, C.G.; Bohlmann, J. Biosynthesis of Sandalwood Oil: *Santalum album* CYP76F cytochromes P450 produce santalols and bergamotol. *PLoS ONE* **2013**, *8*, e75053. [CrossRef] [PubMed]
12. Zhang, X.; Berkowitz, O.; Teixeira da Silva, J.A.; Zhang, M.; Ma, G.; Whelan, J.; Duan, J. RNA-Seq analysis identifies key genes associated with haustorial development in the root hemiparasite *Santalum album*. *Front. Plant Sci.* **2015**, *6*, 661. [CrossRef] [PubMed]
13. Meng, S.; Ma, H.; Wang, S.; Li, Z.; Zhao, Z.; Lu, J. Cloning and the Haustorium-Inducing Factor Response Analysis of *SaRbohA* Gene in *Santalum album Linn*. *Acta Bot. -Occident. Sin.* **2019**, *39*, 2132–2137.
14. Gu, Y.; Wang, Z.; Yang, Z. ROP/RAC GTPase: An old new master regulator for plant signaling. *Curr. Opin. Plant Biol.* **2004**, *7*, 527–536. [CrossRef] [PubMed]
15. Zheng, Z.L.; Yang, Z. The Rop GTPase: An emerging signaling switch in plants. *Plant Mol. Biol.* **2000**, *44*, 1–9. [CrossRef] [PubMed]
16. Yang, Z.; Watson, J.C. Molecular cloning and characterization of rho, a ras-related small GTP-binding protein from the garden pea. *Proc. Natl. Acad. Sci. USA* **1993**, *90*, 8732–8736. [CrossRef]
17. Delmer, D.P.; Pear, J.R.; Andrawis, A.; Stalker, D.M. Genes encoding small GTP-binding proteins analogous to mammalian rac are preferentially expressed in developing cotton fibers. *Mol. Gen. Genet.* **1995**, *248*, 43–51. [CrossRef]
18. Xia, G.; Ramachandran, S.; Hong, Y.; Chan, Y.S.; Simanis, V.; Chua, N.H. Identification of plant cytoskeletal, cell cycle-related and polarity-related proteins using Schizosaccharomyces pombe. *Plant J.* **1996**, *10*, 761–769. [CrossRef]
19. Li, H.; Wu, G.; Ware, D.; Davis, K.R.; Yang, Z. Arabidopsis Rho-related GTPases: Differential gene expression in pollen and polar localization in fission yeast. *Plant Physiol.* **1998**, *118*, 407–417. [CrossRef]
20. Fu, Y.; Li, H.; Yang, Z. The ROP2 GTPase controls the formation of cortical fine F-actin and the early phase of directional cell expansion during *Arabidopsis* organogenesis. *Plant Cell* **2002**, *14*, 777–794. [CrossRef]
21. Ivanchenko, M.; Vejlupkova, Z.; Quatrano, R.S.; Fowler, J.E. Maize ROP7 GTPase contains a unique, CaaX box-independent plasma membrane targeting signal. *Plant J.* **2000**, *24*, 79–90. [CrossRef]

22. Kost, B.; Lemichez, E.; Spielhofer, P.; Hong, Y.; Tolias, K.; Carpenter, C.; Chua, N.H. Rac homologues and compartmentalized phosphatidylinositol 4, 5-bisphosphate act in a common pathway to regulate polar pollen tube growth. *J. Cell Biol.* **1999**, *145*, 317–330. [CrossRef] [PubMed]
23. Lavy, M.; Bracha-Drori, K.; Sternberg, H.; Yalovsky, S. A cell-specific, prenylation-independent mechanism regulates targeting of type II RACs. *Plant Cell* **2002**, *14*, 2431–2450. [CrossRef]
24. Christensen, T.M.; Vejlupkova, Z.; Sharma, Y.K.; Arthur, K.M.; Spatafora, J.W.; Albright, C.A.; Meeley, R.B.; Duvick, J.P.; Quatrano, R.S.; Fowler, J.E. Conserved subgroups and developmental regulation in the monocot *rop* gene family. *Plant Physiol.* **2003**, *133*, 1791–1808. [CrossRef]
25. Winge, P.; Brembu, T.; Bones, A.M. Cloning and characterization of rac-like cDNAs from *Arabidopsis thaliana*. *Plant Mol. Biol.* **1997**, *35*, 483–495. [CrossRef] [PubMed]
26. Bloch, D.; Lavy, M.; Efrat, Y.; Efroni, I.; Bracha-Drori, K.; Abu-Abied, M.; Sadot, E.; Yalovsky, S. Ectopic expression of an activated *RAC* in *Arabidopsis* disrupts membrane cycling. *Mol. Biol. Cell* **2005**, *16*, 1913–1927. [CrossRef]
27. Etienne-Manneville, S.; Hall, A. Rho GTPases in cell biology. *Nature* **2002**, *420*, 629–635. [CrossRef]
28. Hall, A. Rho GTPases and the actin cytoskeleton. *Science* **1998**, *279*, 509–514. [CrossRef]
29. Nibau, C.; Wu, H.M.; Cheung, A.Y. RAC/ROP GTPases: 'hubs' for signal integration and diversification in plants. *Trends Plant Sci.* **2006**, *11*, 309–315. [CrossRef]
30. Yang, Z.; Fu, Y. ROP/RAC GTPase signaling. *Curr. Opin. Plant Biol.* **2007**, *10*, 490–494. [CrossRef] [PubMed]
31. Bokoch, G.M. Regulation of the human neutrophil NADPH oxidase by the Rac GTP-binding proteins. *Curr. Opin. Cell Biol.* **1994**, *6*, 212–218. [CrossRef]
32. Brembu, T.; Winge, P.; Bones, A.M.; Yang, Z. A RHOse by any other name: A comparative analysis of animal and plant Rho GTPases. *Cell Res.* **2006**, *16*, 435–445. [CrossRef] [PubMed]
33. Berken, A. ROPs in the spotlight of plant signal transduction. *Cell Mol. Life Sci.* **2006**, *63*, 2446–2459. [CrossRef] [PubMed]
34. Li, H.; Shen, J.J.; Zheng, Z.L.; Lin, Y.; Yang, Z. The Rop GTPase switch controls multiple developmental processes in Arabidopsis. *Plant Physiol.* **2001**, *126*, 670–684. [CrossRef]
35. Potikha, T.S.; Collins, C.C.; Johnson, D.I.; Delmer, D.P.; Levine, A. The involvement of hydrogen peroxide in the differentiation of secondary walls in cotton fibers. *Plant Physiol.* **1999**, *119*, 849–858. [CrossRef]
36. Park, J.; Choi, H.J.; Lee, S.; Lee, T.; Yang, Z.; Lee, Y. Rac-related GTP-binding protein in elicitor-induced reactive oxygen generation by suspension-cultured soybean cells. *Plant Physiol.* **2000**, *124*, 725–732. [CrossRef]
37. Winge, P.; Brembu, T.; Kristensen, R.; Bones, A.M. Genetic structure and evolution of RAC-GTPases in *Arabidopsis thaliana*. *Genetics* **2000**, *156*, 1959–1971.
38. Miki, D.; Itoh, R.; Shimamoto, K. RNA silencing of single and multiple members in a gene family of rice. *Plant Physiol.* **2005**, *138*, 1903–1913. [CrossRef]
39. Lemichez, E.; Wu, Y.; Sanchez, J.P.; Mettouchi, A.; Mathur, J.; Chua, N.H. Inactivation of *AtRac1* by abscisic acid is essential for stomatal closure. *Genes Dev.* **2001**, *15*, 1808–1816. [CrossRef] [PubMed]
40. Zheng, Z.L.; Nafisi, M.; Tam, A.; Li, H.; Crowell, D.N.; Chary, S.N.; Schroeder, J.I.; Shen, J.; Yang, Z. Plasma membrane-associated ROP10 small GTPase is a specific negative regulator of abscisic acid responses in *Arabidopsis*. *Plant Cell* **2002**, *14*, 2787–2797. [CrossRef]
41. Jones, M.A.; Shen, J.J.; Fu, Y.; Li, H.; Yang, Z.; Grierson, C.S. The Arabidopsis Rop2 GTPase is a positive regulator of both root hair initiation and tip growth. *Plant Cell* **2002**, *14*, 763–776. [CrossRef]
42. Chen, L.; Shiotani, K.; Togashi, T.; Miki, D.; Aoyama, M.; Wong, H.L.; Kawasaki, T.; Shimamoto, K. Analysis of the Rac/Rop small GTPase family in rice: Expression, subcellular localization and role in disease resistance. *Plant Cell Physiol.* **2010**, *51*, 585–595. [CrossRef] [PubMed]
43. Ono, E.; Wong, H.L.; Kawasaki, T.; Hasegawa, M.; Kodama, O.; Shimamoto, K. Essential role of the small GTPase *Rac* in disease resistance of rice. *Proc. Natl. Acad. Sci. USA* **2001**, *98*, 759–764. [CrossRef]
44. Jung, Y.H.; Agrawal, G.K.; Rakwal, R.; Kim, J.A.; Lee, M.O.; Choi, P.G.; Kim, Y.J.; Kim, M.J.; Shibato, J.; Kim, S.H.; et al. Functional characterization of *OsRacB* GTPase–a potentially negative regulator of basal disease resistance in rice. *Plant Physiol. Biochem.* **2006**, *44*, 68–77. [CrossRef]
45. Yang, Z. Small GTPases: Versatile signaling switches in plants. *Plant Cell* **2002**, *14*, 375–388. [CrossRef] [PubMed]
46. Lin, Y.; Zeng, Y.; Zhu, Y.; Shen, J.; Ye, H.; Jiang, L. Plant Rho GTPase signaling promotes autophagy. *Mol. Plant* **2021**, *14*, 905–920. [CrossRef]
47. Li, H.; Lin, Y.; Heath, R.M.; Zhu, M.X.; Yang, Z. Control of pollen tube tip growth by a Rop GTPase-dependent pathway that leads to tip-localized calcium influx. *Plant Cell* **1999**, *11*, 1731–1742.
48. Molendijk, A.J.; Bischoff, F.; Rajendrakumar, C.S.; Friml, J.; Braun, M.; Gilroy, S.; Palme, K. *Arabidopsis thaliana* Rop GTPases are localized to tips of root hairs and control polar growth. *EMBO J.* **2001**, *20*, 2779–2788. [CrossRef] [PubMed]
49. Baxter-Burrell, A.; Yang, Z.; Springer, P.S.; Bailey-Serres, J. RopGAP4-dependent Rop GTPase rheostat control of *Arabidopsis* oxygen deprivation tolerance. *Science* **2002**, *296*, 2026–2028. [CrossRef]
50. Torres, M.A.; Dangl, J.L.; Jones, J.D. *Arabidopsis* gp91phox homologues AtrbohD and AtrbohF are required for accumulation of reactive oxygen intermediates in the plant defense response. *Proc. Natl. Acad. Sci. USA* **2002**, *99*, 517–522. [CrossRef]

51. Foreman, J.; Demidchik, V.; Bothwell, J.H.; Mylona, P.; Miedema, H.; Torres, M.A.; Linstead, P.; Costa, S.; Brownlee, C.; Jones, J.D.; et al. Reactive oxygen species produced by NADPH oxidase regulate plant cell growth. *Nature* **2003**, *422*, 442–446. [CrossRef] [PubMed]
52. Kwak, J.M.; Mori, I.C.; Pei, Z.M.; Leonhardt, N.; Torres, M.A.; Dangl, J.L.; Bloom, R.E.; Bodde, S.; Jones, J.D.; Schroeder, J.I. NADPH oxidase *AtrbohD* and *AtrbohF* genes function in ROS-dependent ABA signaling in *Arabidopsis*. *EMBO J.* **2003**, *22*, 2623–2633. [CrossRef] [PubMed]
53. Yoshioka, H.; Numata, N.; Nakajima, K.; Katou, S.; Kawakita, K.; Rowland, O.; Jones, J.D.; Doke, N. Nicotiana benthamiana gp91phox homologs NbrbohA and NbrbohB participate in H2O2 accumulation and resistance to Phytophthora infestans. *Plant Cel.* **2003**, *15*, 706–718. [CrossRef] [PubMed]
54. Jones, M.A.; Raymond, M.J.; Yang, Z.; Smirnoff, N. NADPH oxidase-dependent reactive oxygen species formation required for root hair growth depends on ROP GTPase. *J. Exp. Bot.* **2007**, *58*, 1261–1270. [CrossRef]
55. Zhang, X.; Teixeira da Silva, J.A.; Duan, J.; Deng, R.; Xu, X.; Ma, G. Endogenous hormone levels and anatomical characters of haustoria in *Santalum album* L. seedlings before and after attachment to the host. *J. Plant Physiol.* **2012**, *169*, 859–866. [CrossRef]
56. Meng, S.; Wang, X.L.; Bian, Z.; Li, Z.S.; Yang, F.C.; Wang, S.K.; John, I.Y.; Lu, J.K. Melatonin enhances nitrogen metabolism and haustorium development in hemiparasite *Santalum album* Linn. *Environ. Exp. Bot.* **2021**, *186*, 104460. [CrossRef]
57. Kim, Y.K.; Kim, Y.B.; Kim, J.K.; Kim, S.U.; Park, S.U. Molecular cloning and characterization of mevalonic acid (MVA) pathway genes and triterpene accumulation in Panax ginseng. *J. Korean Soc. Appl. Biol.* **2014**, *57*, 289–295. [CrossRef]
58. Kim, T.H.; Ito, H.; Hayashi, K.; Hasegawa, T.; Machiguchi, T.; Yoshida, T. Aromatic constituents from the heartwood of *Santalum album* L. *Chem. Pharm. Bull. (Tokyo)* **2005**, *53*, 641–644. [CrossRef] [PubMed]
59. Luan, Y.; Wang, B.; Zhao, Q.; Ao, G.; Yu, J. Ectopic expression of foxtail millet zip-like gene, SiPf40, in transgenic rice plants causes a pleiotropic phenotype affecting tillering, vascular distribution and root development. *Sci. China Life Sci.* **2010**, *53*, 1450–1458. [CrossRef]

life MDPI

Article

The Molecular Network behind Volatile Aroma Formation in Pear (*Pyrus* spp. Panguxiang) Revealed by Transcriptome Profiling via Fatty Acid Metabolic Pathways

Huiyun Li [1,†], Jine Quan [1,†], Sohel Rana [1,†], Yanmei Wang [1], Zhi Li [1], Qifei Cai [1], Shuhong Ma [2], Xiaodong Geng [1,*] and Zhen Liu [1,*]

1 College of Forestry, Henan Agricultural University, Zhengzhou 450002, China
2 Ying Lin Station Anyang Forestry Bureau, Anyang 455000, China
* Correspondence: xiaodonggeng@163.com (X.G.); liuzhen@henau.edu.cn (Z.L.); Tel.: +86-150-9315-9899 (X.G.); +86-138-3808-8526 (Z.L.)
† These authors contributed equally to this work.

Simple Summary: Pear is a widely eaten fruit all over the world. Volatile aroma is an important factor affecting fruit quality and the fatty acid metabolism pathway is important in synthesizing volatile aromas. In this study, Panguxiang (*Pyrus* spp. Panguxiang) is a new variety bred from *P. bretschneideri* Rehd. cv. 'Biyang piaoli' and, unlike most white pear varieties cultivated in China, its aroma is also vital. The study aimed to explore unique pear resources of rich fruit aroma and to clarify the metabolism and regulation mechanism of the aromatic components in pear fruit. This paper used physiological and transcriptome methods to explore the molecular network behind volatile aroma formation in Panguxiang revealed via fatty acid metabolic pathways. Through transcriptome sequencing, weighted gene co-expression network analysis (WGCNA) identified yellow functional modules and several biological and metabolic pathways related to fatty acid formation. Finally, we identified seven and eight hub genes in the fatty acid synthesis and fatty acid metabolism pathways, respectively. Further analysis of the co-expression network allowed us to identify several key transcription factors related to the volatile aroma, including AP2/ERF-ERF, C3H, MYB, NAC, C2H2, GRAS, and Trihelix, which may also be involved in fatty acid synthesis and further influence the formation of aroma.

Abstract: Pears are popular table fruits, grown and consumed worldwide for their excellent color, aroma, and taste. Volatile aroma is an important factor affecting fruit quality, and the fatty acid metabolism pathway is important in synthesizing volatile aromas. Most of the white pear varieties cultivated in China are not strongly scented, which significantly affects their overall quality. Panguxiang is a white pear cultivar, but its aroma has unique components and is strong. The study of the mechanisms by which aroma is formed in Panguxiang is, therefore, essential to improving the quality of the fruit. The study analyzed physiological and transcriptome factors to reveal the molecular network behind volatile aroma formation in Panguxiang. The samples of Panguxiang fruit were collected in two (fruit development at 60, 90, 120, and 147 days, and fruit storage at 0, 7, 14, 21, and 28 days) periods. A total of nine sample stages were used for RNA extraction and paired-end sequencing. In addition, RNA quantification and qualification, library preparation and sequencing, data analysis and gene annotation, gene co-expression network analysis, and validation of DEGs through quantitative real-time PCR (qRT-;PCR) were performed in this study. The WGCNA identified yellow functional modules and several biological and metabolic pathways related to fatty acid formation. Finally, we identified seven and eight hub genes in the fatty acid synthesis and fatty acid metabolism pathways, respectively. Further analysis of the co-expression network allowed us to identify several key transcription factors related to the volatile aroma, including AP2/ERF-ERF, C3H, MYB, NAC, C2H2, GRAS, and Trihelix, which may also be involved in the fatty acid synthesis. This study lays a theoretical foundation for studying volatile compounds in pear fruits and provides a theoretical basis for related research in other fruits.

Citation: Li, H.; Quan, J.; Rana, S.; Wang, Y.; Li, Z.; Cai, Q.; Ma, S.; Geng, X.; Liu, Z. The Molecular Network behind Volatile Aroma Formation in Pear (*Pyrus* spp. Panguxiang) Revealed by Transcriptome Profiling via Fatty Acid Metabolic Pathways. *Life* **2022**, *12*, 1494. https://doi.org/10.3390/life12101494

Academic Editor: Othmane Merah

Received: 28 August 2022
Accepted: 21 September 2022
Published: 26 September 2022

Keywords: pears; volatile aroma formation; fatty acid pathway; RNA-Seq; WGCNA

1. Introduction

Aroma is an important indicator of fruit quality [1], and rich fruit aromas can alter trends in consumption, improving human standards of living and increasing demand for fruits. The study of fruit aroma is therefore receiving increasing attention [2].

The types and proportions of the molecular components of fruit aroma can differ significantly between species, and there are also differences between different varieties of the same species. Pears are a vital fruit globally, with white pears (*Pyrus bretschneideri*) being the most widely cultivated varieties in China, therefore being often referred to as "Chinese pears". Panguxiang (*Pyrus* spp. Panguxiang) is a new variety bred from *P. bretschneideri* Rehd. cv. 'Biyang piaoli'. Panguxiang is highly nutritious and tastes sweet, crisp, and slightly sour. The aroma has unique components, but unlike most white pear varieties cultivated in China, the aroma is also strong. Due to scattered operations and poor management by farmers, the quality of Panguxiang has gradually decreased. Therefore, it is of great significance to study the anabolic mechanism behind the aroma of Panguxiang and effectively promote the regulation and accumulation of volatile substances.

Research into the volatile substances produced by pears began in 1964 [3]. Since then, there have been many studies on the characteristic volatiles of different pear species or varieties [4,5], including on the synthesis of aromatic compounds [6] and on the influence of ripening [7] and storage [8] stages on the aroma. Researchers have detected more than 300 volatile substances released by pear fruits, including aldehydes, alcohols, esters, terpenes, hydrocarbons, and sulfur-containing compounds [9]. Of these compounds, volatile esters and alcohols are known to be the most significant contributors to the aroma of certain fruits, including apple [10], strawberry [11], and banana [12]. A previous study reported that fatty acid metabolism is a major pathway through which volatiles (particularly esters) are formed in pears following the incubation of fruits with linoleic acid (LA, C18:2) and linolenic acid (LNA, C18:3) [13].

Straight-chain aliphatic alcohols, aldehydes, ketones, and esters are mainly generated via the fatty acid metabolic pathway [14]. In this pathway, the synthesis of volatile substances begins with C18 organic compounds, such as unsaturated fatty acids. The LA and LNA can generate straight-chain aldehydes, alcohols, and esters through the lipoxygenase (LOX) enzymatic system, which comprises four key enzymes, including LOX, hydroperoxide lyase (HPL), alcohol dehydrogenase (ADH), and alcohol acyltransferase (AAT) [15]. After LA and LNA enter the LOX pathway, they first form two intermediates—9-hydroperoxide and 13-hydroperoxide—under the catalysis of LOX, which are then further metabolized through the LOX pathway [16]. Next, these two hydroperoxide fatty acid derivatives are converted into C6 and C9 aldehydes through the action of HPL. These aldehydes are reduced to alcohols under the action of ADH and are further converted into esters under the action of AAT [14]. The ester compounds catalyzed by AAT are responsible for the strong aromas of fruits.

There have been many studies on volatile compounds in different pear varieties [5,17–20]. However, limited research has concentrated on the mechanism of volatile compound synthesis in Panguxiang. The current study focuses on the changes in volatile substance synthesis during the development and storage of Panguxiang fruits. In addition, the molecular regulatory network of volatile substance synthesis during fruit ripening and storage was analyzed with the WGCNA package. The results lay a theoretical foundation for the mechanism of fruit aroma synthesis and provide a theoretical basis for improving Panguxiang aroma.

2. Materials and Methods

2.1. Plant Materials and Sample Preparation

This experiment was carried out in the Key Laboratory of Forest Resources Cultivation of Henan Agricultural University and the State Forestry Administration (112°42′114°14′ E, and 34°16′34°58′ N). The soil condition at the experimental site was sandy loam with a pH value of 7.0. The study was conducted in 2021. A 6-year-old Panguxiang pear tree with strong growth was selected as the test tree. Thirty days after full flowering (May 2021), the middle and upper parts of the fruit tree in the four cardinal directions without diseases and pests were selected, and uniform fruits were produced for bagging.

To analyze the variations of the parameters in different stages (growth and development and storage period), a total of nine stages were selected to sample the fruits. Four stages of pear fruit growth and development were sampled: 60 days (S1), 90 days (S2), 120 days (S3), and 147 days (S4) after anthesis. When the fruits on the tree were ripe, samples of similar size were picked and stored in an artificial incubator (temperature 20 °C, humidity 70%) as samples for the storage period. Five stages of pear fruit storage were sampled: 0 days (Z1), 7 days (Z2), 14 days (Z3), 21 days (Z4), and 28 days (Z5) of storage. After sampling, all samples were stored in a −80 °C refrigerator. All samples at the nine stages were used for the RNA extraction and paired-end sequencing (samples from each period were collected in triplicate).

2.2. Determination of Volatile Substances

The samples taken at the nine different stages were removed from the −80 °C storage freezer and separately ground in liquid nitrogen. The samples taken at each stage were evenly divided into three 15 g replicates, and a total of 27 samples were used to determine the content of volatile substances. The specific test method follows that described by Fan et al. [21].

2.3. RNA Quantification and Qualification

Total RNA was isolated and purified using the CTAB method [22]. The integrity, purity, and concentration of the purified RNA were assessed using an Agilent 2100 Bioanalyzer and a NanoDrop ND-1000 spectrophotometer (NanoDrop Technologies, Wilmington, DE, USA). The mRNA extracted from the total RNA in the samples was isolated using Oligo dT.

2.4. Library Preparation and Sequencing

Libraries were generated and purified using the NEBNext® Ultra™ RNA Library Prep Kit for Illumina® (New England Biolabs Inc., Ipswich, MA, USA) and AMPure XP Beads (Beckman Coulter, Inc., Indianapolis, IN, USA), using the fragmented mRNA as the template and following the manufacturer's recommendations. The concentration, integrity, and quantification of the library were determined by using a Qubit™ Fluorometer (ThermoFisher Scientific, Waltham, MA, USA), the KAPA Library Quantification Kit (KAPA Biosystems, Wilmington, MA, USA), and a Qsep100 DNA Analyzer (KAPA Biosystems), respectively. The denatured libraries were subjected to high-throughput parallel sequencing of both ends of the library using an Illumina HiSeq X™ Ten System sequencing platform. The quality of the raw data was evaluated using default settings.

2.5. Data Analysis and Gene Annotation

The raw data were filtered through fastp v0.19.3 [23] to obtain clean data. HISAT v2.1.0 [24] was used to map clean reads to the reference genome. The transcripts were quantified using featureCounts v1.6.2 [25], and the lengths of the transcripts in the sample were normalized to FPKM (fragments per kilobase of exon per million fragments mapped) values. The differential gene expression was analyzed between different groups using DESeq2 v1.22.1 [26]. After the difference analysis, the p value was corrected using the Benjamini and Hochberg method to obtain the false discovery rate (FDR). Genes with an expression-level change of $|\log2 \text{ (fold change)}| \geq 1$ and FDR < 0.05 were considered to be differentially expressed genes (DEGs). The transcript sequences were mapped to seven public databases: the NCBI non-redundant protein sequences database (NR) (https://www.ncbi.nlm.nih.gov/ accessed on 25 August 2022) [27]; Swiss-Prot [28,29]; Gene Ontology (GO) [30]; euKaryotic Ortholog Groups (KOG) [31]; Protein family (Pfam) [32]; Kyoto Encyclopedia of Genes and Genomes (KEGG) [33]; and TrEMBL [34]. Mapping was conducted using BLAST software (E-value $\leq 10^{-5}$) to obtain annotation information for the transcripts [35]. The protein sequences were submitted to iTAK [36] to identify the transcription factors.

2.6. Gene Co-Expression Network Analysis

All the identified DEGs were used to construct a co-expression network using the R package WGCNA [37]. The co-expression modules were obtained using the automatic network construction function (blockwiseModules) with power = 18 and minModuleSize = 60. The TOM type was unsigned and the modules with highly correlated eigengenes (correlation > 0.85) were merged. The eigenvalue was calculated for each module based on Pearson correlation. The networks were visualized by Cytoscape [38].

2.7. Validation of DEGs through Quantitative Real-Time PCR (qRT–PCR)

Total RNA extracted from the fruit of Panguxiang was reverse transcribed using the FastQuant RT Kit with DNase (TianGen Biotech Co., Ltd., Beijing, China) to synthesize first-strand cDNA. A qRT–PCR assay was performed with an optical 96-well reaction plate, the ABI PRISM 7500 Real-time PCR system (Applied Biosystems, Foster City, CA, USA), and SuperReal PreMix Plus SYBR Green (TianGen Biotech Co., Ltd.). Each reaction contained 12.5 µL SYBR Premix ExTaq, 0.5 µL ROX reference dye, 2.0 µL cDNA, and 1.0 µL gene-specific primers in a final volume of 25 µL. All the primers were at concentrations of 10 µM. The primer sequence of genes is presented in Table 1. The PCR program was as follows: 95 °C for 10 s and then 45 cycles at 95 °C for 5 s and 60 °C for 40 s; tubulin was used as a reference. The qRT–PCR data were analyzed using the $2^{-\Delta\Delta CT}$ method [39]. The RNA concentration ranged from 600 ng/µL to 800 ng/µL, and the A260/A280 value ranged from 1.8 to 2.0. The initial concentration of cDNA was 1 µg/µL. The cDNA was diluted to four concentrations: 10^0, 10^{-1}, 10^{-2}, and 10^{-3} (1 µg/µL). The E-value (amplification efficiency) of the qRT–PCR was between 90% and 110%, and R^2 was greater than 0.99. The qRT–PCR of each gene was carried out three times to give three experimental replicates, and each experiment comprised three biological replicates [40].

Table 1. The primer sequence of genes.

Gene Name	Gene ID	Primer Sequence (5′-3′)		The Size Products
long-chain acyl-CoA synthetase	LOC103949217	F:AACCGACTATCTTCTGTGCTGTTCC	R:AGCCGCCTGATGAAATCTTCTGTG	81 bp
lysophospholipid acyltransferase	LOC103931841	F:CCAATGCTACTCGGCTATGCTTCC	R:CGATTCCTCCTTCCTTCCATGCG	139 bp
alcohol dehydrogenase (NADP+)	LOC103952287	F:CACCAGCATTGGGCTTGAAATCG	R:CACCTCCATCCCTGCCTCCTTC	121 bp
endogenous reference gene	AY338250	F:CAGGCTGACTGTGCTGTCCT	R:TCACACCGAGGGTGAAAGCA	115 bp

3. Results

3.1. Sampling and Determination of Volatile Substance Content

We measured the fruit weight and size at the time of sampling. With the ripening of the fruits, their weight and size also increased, with the weight at stage S4 reaching 260 g. (Figure 1a–d). To analyze the aroma components of Panguxiang fruits at different stages, we determined the volatile substances present in the nine stages of the fruits. These volatiles included alcohols, ketones, aldehydes, esters, and terpenes. The levels of alcohols, ketones, and aldehydes in the nine stages initially decreased with ripening, but increased significantly starting at stage S3. The concentrations of aldehydes and ketones continued to increase throughout the first storage stages, reaching maxima at Z3 and Z4, after which they decreased once more. In contrast, the concentrations of alcohols, although lower than those of the other chemical groups, continued to rise throughout storage. The concentrations of esters and terpenoids remained unchanged throughout the ripening stages, but increased sharply during the first three stages of storage to a maximum at stage Z3 (4265 ng. g^{-1} and 4394 ng. g^{-1}, respectively), after which they decreased again. At the early stage of storage, from Z1 to Z3, fruit esters and terpenoids rose rapidly; at the later stage of storage, the decrease rate of ester concentrations was slightly slower than that of terpenoid concentrations (Figure 1e,f).

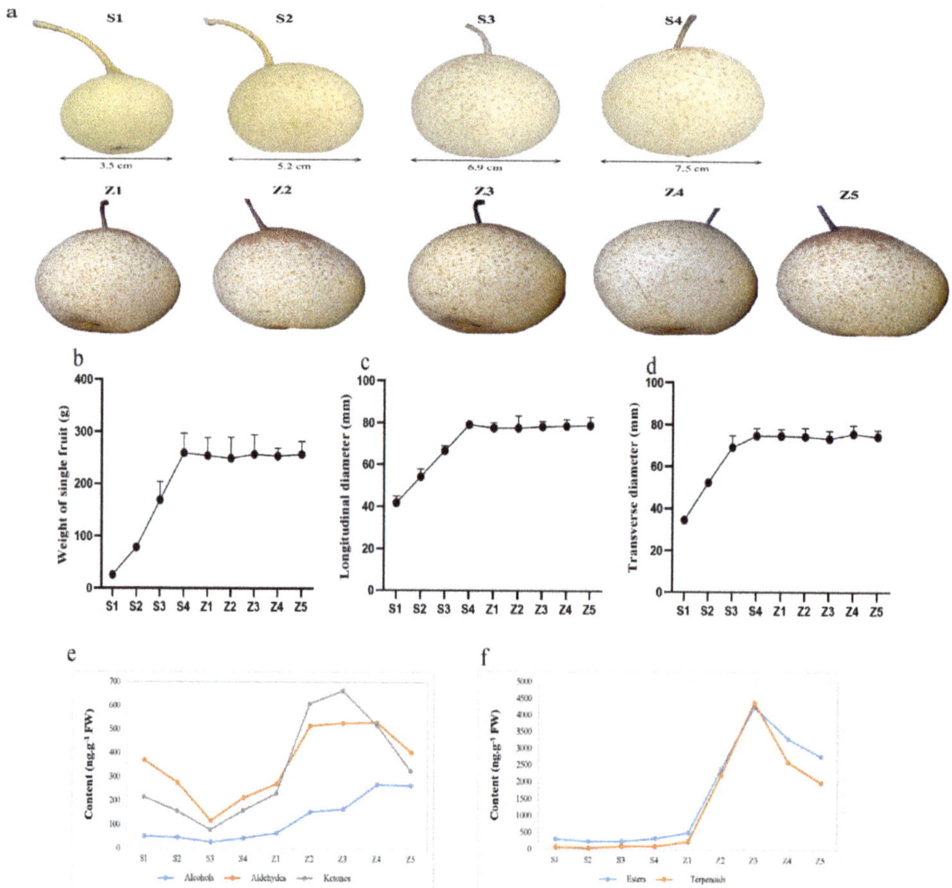

Figure 1. Phenotypic analysis of Panguxiang in different ripening and storage stages. (**a**) The phenotype of Panguxiang over the nine studied stages. (**b–d**) Visualized phenotypic analysis of ripening fruits. (**e**,**f**) Determination of concentrations of volatile compound groups in fruit over the nine studied stages.

3.2. Transcriptomic Analysis of *Pyrus spp*. Panguxiang Fruits

This study investigated the transcriptomic sequences of 27 samples taken at different fruit ripening and storage stages. A total of 183.32 GB of clean data was obtained. The clean data from each sample reached 6 GB and the Q30 score was >91% in each case (Table S1). The clean read sequences were aligned to the reference genome using HISAT2. The overall alignment ratio of each sample was higher than 70% (Table S2). The FPKM distribution of the transcriptome sequence data was visualized with box plots and violin plots to compare the overall transcript expression levels in the different samples. Gene expression was stable between the 27 samples (Figure S1a,b). Principal component analysis (PCA) and Pearson correlation coefficient (PCC) confirmed that the transcriptome characteristics were highly correlated between the biological replicates of each group of samples (Figure S1c,d).

A BLAST search was conducted to explore the functions of the unigenes and obtain annotation information for the transcripts. Functional annotations were performed using multiple public databases, including Nr, KEGG, GO, SwissProt, KOG, TrEMBL, and Pfam. The numbers of transcripts annotated by the seven databases were as follows: 39,707 in Nr, 34,069 in GO, 29,399 in KEGG, 24,306 in KOG, 33,220 in Pfam, 29,449 in SwissProt, and 40,424 in TrEMBL (Table S3). In identifying the DEGs in the nine different study stages, RNA sequencing (RNA-Seq) was performed to profile the dynamic changes in genome-wide transcript abundance and pairwise DESeq2 analysis between different time points was carried out. Hierarchical cluster analysis was then conducted according to the FPKM expression data after the standardization of differential genes, and cluster heatmaps were drawn (Figure S1e). We found large and significant differences in the number of DEGs among the nine study stages. In the comparative analysis of DEGs in different storage periods, the number of DEGs decreased from Z2/Z1 4382 (2638 upregulated and 1744 downregulated) to the subsequent period Z3/Z2 1184 (635 upregulated and 549 downregulated), respectively (Table S4). There were 1137 DEGs in Z4/Z3 (770 upregulated and 367 downregulated) and 1917 DEGs in Z5/Z4 (1285 upregulated and 632 downregulated). In the early stage of fruit storage, most DEGs were found in Z2/Z1. It is also possible that the transcriptional expression level of genes precedes the phenotypic changes observed (Figure S1f, Table S2).

3.3. Functional Annotation and Classification of DEGs

We analyzed the GO, KOG, and KEGG pathways to elucidate the biological functions of DEGs. The GO annotation system consists of three major branches: "biological process", "molecular function", and "cellular component". These unigenes were further divided into 49 to 57 major functional terms. The "cellular process", "cell part", and "binding" were the most over-represented terms in the three GO categories mentioned above (Figure S2). The unigenes enriched by KOG could be assigned to 25 groups. Group R (general function prediction only) was the most highly represented. Groups T (signal transduction mechanisms) and O (posttranslational modification, protein turnover, chaperones) also shared a high percentage of genes. Only a few genes were assigned to Groups Y (nuclear structure), W (extracellular structures), and N (cell motility) (Figure S3). The "fatty acid metabolism pathway" was also indicated in the KEGG signaling pathway (Figure S4).

3.4. Co-Expression Analysis Identifies Key Gene Modules Involved in the Formation of Aroma Volatiles

To fully understand the correlation of gene expression during the formation of aroma volatiles, all DEGs were used in WGCNA and 15 co-expressed gene modules were identified (Figure S5). Module size ranged from 87 to 11,092 genes, with black (721 genes), blue (3377 genes), brown (3111 genes), cyan (136 genes), green (1519 genes), green-yellow (350 genes), grey (287), magenta (625 genes), midnight blue (87 genes), pink (662 genes), purple (613 genes), red (827 genes), salmon (263 genes), tan (346 genes), yellow (1953 genes), and turquoise (11,092 genes) represented.

The gene expression profile was visualized with the eigengene values for each module and showed distinct co-expression patterns across modules (Figure S5). The expression trend of the yellow and purple modules is consistent with the changing trend of aroma components in pear fruit. None of the other 13 modules were correlated with any variable. These results suggest that the yellow and purple modules may be the key to understanding the mechanism of pear aroma formation. However, the network formed by key genes in the purple module was less extensive, so we focused here on the analysis of the yellow module.

Transcription factors (TFs) are important regulators in plant development [41], and we therefore investigated the distribution of TF genes in all modules (Table S5). Statistical analysis suggested that the prominent TF families present in the yellow module were the AP2/ERF-ERF, C3H, MYB, NAC, C2H2, GRAS, and Trihelix families (Table 2, Table S6).

Table 2. The number and types of TFs in different modules.

TF Family	Black	Blue	Brown	Cyan	Green	Green-Yellow	Grey	Magenta	Midnight Blue	Pink	Purple	Red	Salmon	Tan	Yellow	Turquoise
AP2/ERF-ERF		21	7	1	10					10			2		8	38
C2H2	1	4	8		5			2		3	2	6		1	5	32
C3H		13	9	1	10		2	2		2		5		1	7	22
GRAS		4	10		10										5	26
MYB	1		7	1	7		1	4		3	2	2			7	37
NAC	2	4	11		8				1	4	2				6	36
Trihelix		2	1		4			1		2					6	22
Total	4	48	53	3	54	0	3	9	1	24	6	13	2	2	44	213

3.5. Co-Expression Network Analysis

To explore the correlations between each module and the samples taken at different stages, we drew a heatmap illustrating the correlations between the samples and the modules. The yellow and purple modules had the strongest correlations with later storage stages (Figure 2a), which is consistent with the previous result. The co-expression network analysis showed correlations between gene expression patterns and we constructed a network of detected co-expressed modules to identify key hub genes. A few hub genes interact with many other genes in gene networks. In this study, we analyzed the co-expression networks of the fatty acid synthesis pathway and the fatty acid metabolism pathway in the yellow module. We identified seven and eight key hub genes, respectively. The important regulatory genes involved in the fatty acid synthesis pathway were identified in those from the alcohol dehydrogenase (ADH 1, lysophospholipid acyltransferase LOC103931841; ADH 2, phospholipase A1LOC103965089; ADH 3, alcohol dehydrogenase (NADP+)LOC103949386; ADH 4, shikimate O-hydroxycinnamoyltransferase LOC103944272; ADH 5, 1-acyl-sn-glycerol-3-phosphate acyltransferaseLOC103938216; and ADH 6, palmitoyltransferase LOC103940564) and alcohol acyltransferase (AAT; alcohol dehydrogenase (NADP+) LOC103952287) genes, and the important regulatory genes involved in the fatty acid metabolism pathway were identified (acetyl-CoA carboxylase, biotin carboxylase subunit LOC103928973; 3-oxoacyl-[acyl-carrier protein] reductaseLOC103944413; long-chain acyl-CoA synthetaseLOC103949217; acetyl-CoA carboxylase biotin carboxyl carrier proteinLOC103950990; enoyl-[acyl-carrier protein] reductaseLOC103953997; long-chain acyl-CoA synthetaseLOC103954143; acetyl-CoA carboxylase biotin carboxyl carrier protein LOC103955683, and novel.2114 (glycerol-3-phosphate acyltransferase)). (Figure 2b,c).

3.6. Verification of the Expression Profiles of Key Genes

To verify the RNA-Seq data, we used qRT–PCR to analyze the expression levels of three genes involved in the modules from the fatty acid synthesis and metabolism pathways. The gene expression patterns were then compared with the FPKM values. The results showed that the expression patterns suggested by the two analysis techniques were consistent, indicating that our data were reliable (Figure 3 and Table 1).

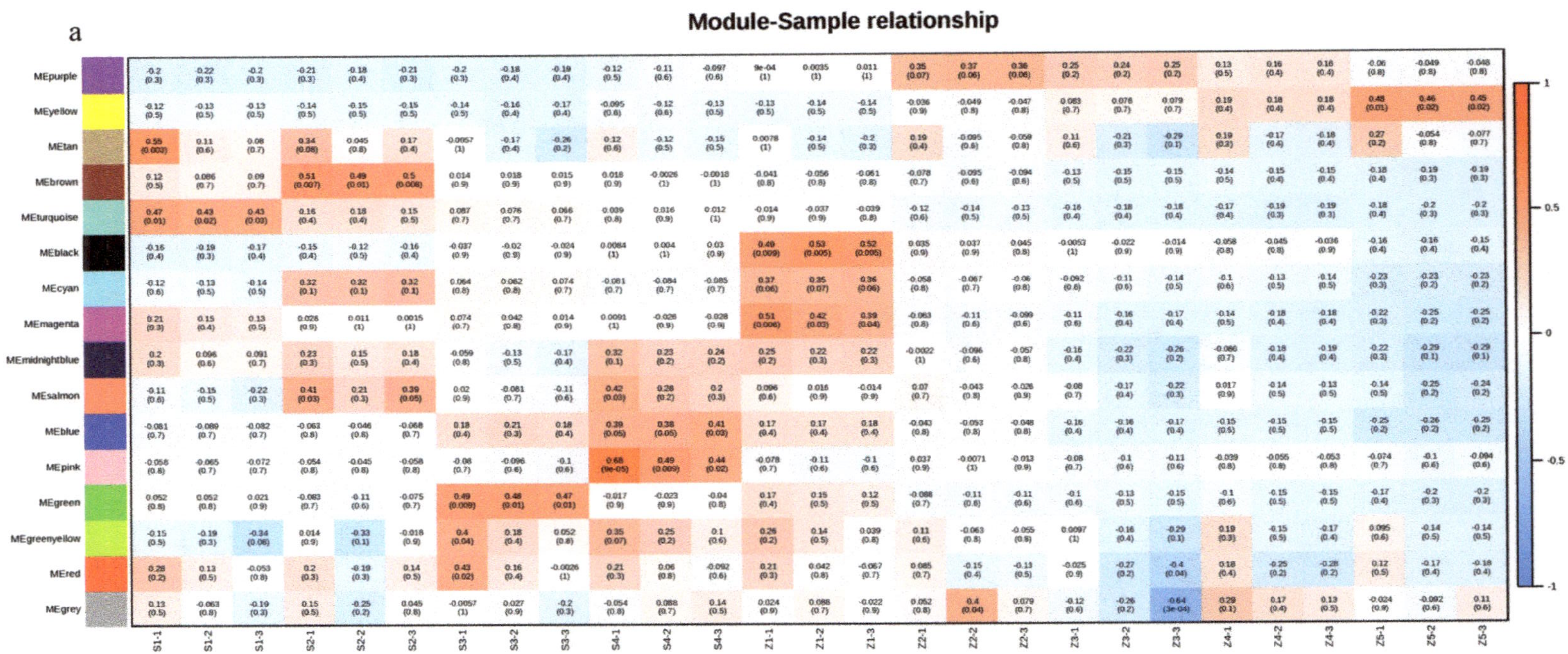

Figure 2. *Cont.*

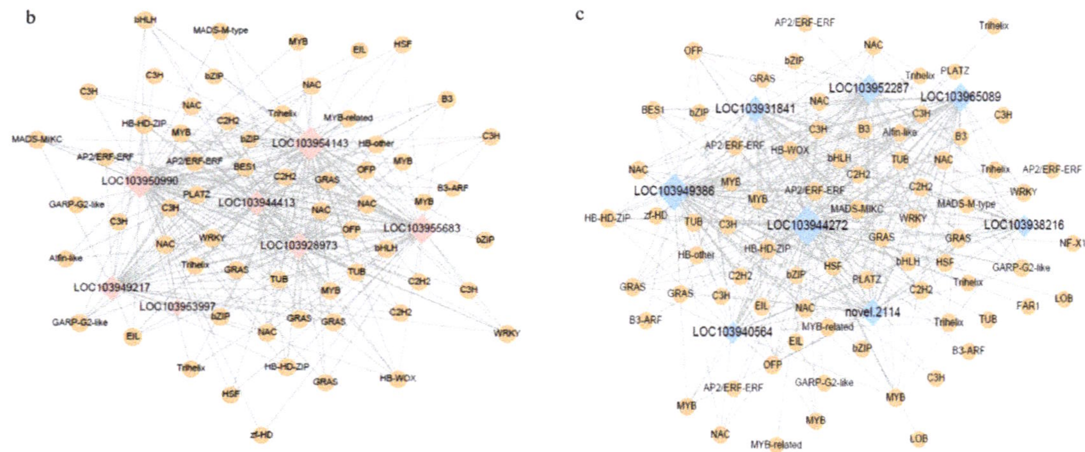

Figure 2. The co-expressed gene modules were identified by weighted gene co-expression network analysis (WGCNA). (**a**) Heatmap of correlations of expressed genes between samples and modules. (**b**) Co-expression network of the fatty acid synthesis pathway in the yellow module. (**c**) Co-expression network of the fatty acid metabolism pathway in the yellow module. For emphasis purposes, the circles representing hub genes are artificially enlarged.

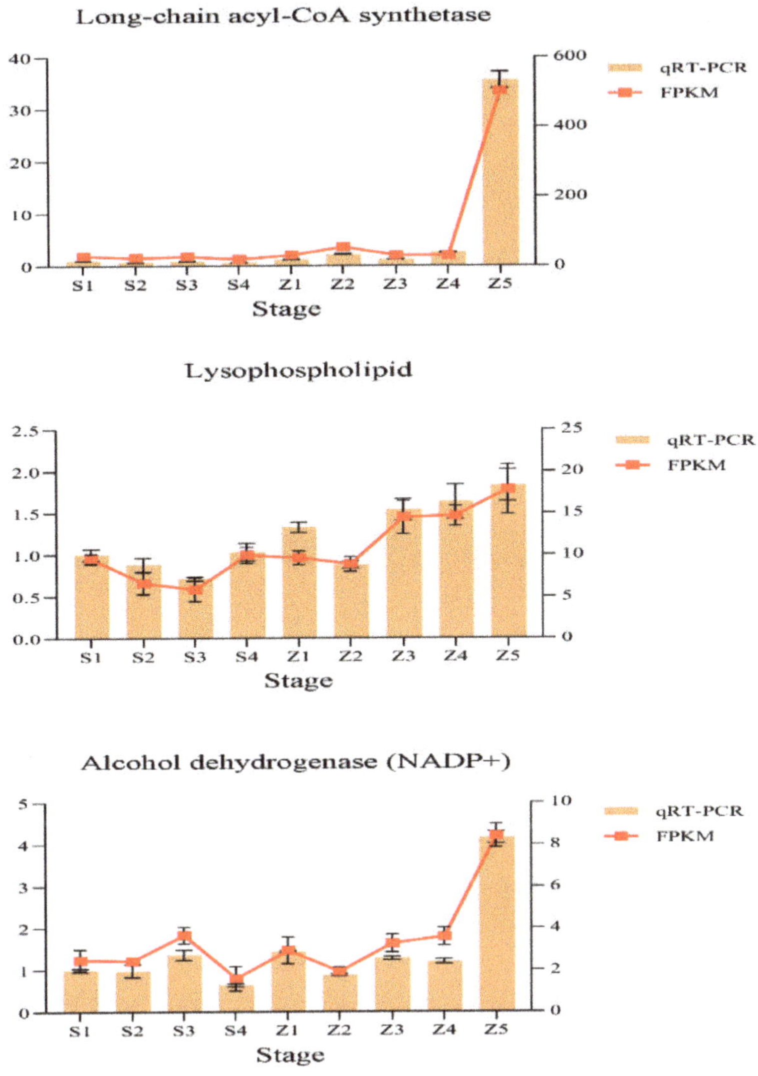

Figure 3. Verification of the expression profiles of representative genes from different modules using quantitative real-time PCR. Relative gene expression levels were calculated using tubulin as the internal reference. The results are presented as the means ± SE, and each sample represents three biological replicates.

4. Discussion

4.1. Effect of Fruit Ripening and Storage on Aroma Content

The synthesis of fruit aroma is a dynamic process and most substances important for fragrance appear in the later stage of fruit development, stimulated by the release of ethylene [42]. Of the different classes of volatile chemicals, research suggests that esters and aldehydes are the most abundant in pears, regardless of species or cultivar [6,43]. Unripe fruits mainly contain "green flavor" aroma substances such as aldehydes and alcohols, with aldehyde and alcohol concentrations gradually decreasing and ester concentrations gradually increasing as the fruits mature [44]. As the most abundant class of volatiles, esters contribute 'sweet' and 'fruity' notes to the aroma of different fruits [45]. This is consistent with our findings. In our study, the size and weight of the fruits increased as they matured, and the concentrations of alcohols, ketones, and aldehydes decreased. At the same time, ester concentrations tended to remain constant, which may result from accumulation. Esters were significantly upregulated when the fruits were fully ripe, with the upregulation lasting until the Z3 stage. The concentrations of volatile aroma components showed a similar trend during storage, increasing first and then decreasing, but during the storage stages, the esters and terpenoids were at much higher levels than the aldehydes and alcohols. It is possible that the fruits continued to ripen, even after picking, until stage Z3. At stage Z3, the volatile aroma content peaked, after which the fruits were affected by the low storage temperature and the concentrations of volatile aroma components began to decrease. In general, the main volatile aromas produced by Panguxiang during fruit growth and development are alcohols and aldehydes, while the main volatile aromas released during storage are esters and terpenoids.

4.2. WGCNA Explores the Formation of Volatile Aromas in Pear Fruits

WGCNA is a widely used data mining method. Exploration of the genes whose expression patterns closely correlate with each other can be achieved through the application of WGCNA [46]. Although WGCNA can be applied to most high-dimensional data sets, it is most widely used in genomes, especially transcriptomes [47]. It allows the definition of a module's network nodes, hub, and membership, as well as the study of the relationship between the co-expressed modules and the comparison of the topologies of different networks [37]. Through this algorithm, the genes that are co-expressed with high correlation are placed in one module, and different modules may contain key genes implicated in the volatile aroma formation process [48].

Our results showed that the purple and yellow modules had the strongest correlation with the S3, S4, and S5 stages, and samples taken at the later storage stages also had the highest volatile aroma content. Based on our choice, the yellow module was also an essential factor. In the yellow module, we analyzed the co-expression networks of two pathways: the fatty acid synthesis pathway and the fatty acid metabolism pathway. We identified seven and eight key hub genes from these two pathways.

4.3. Key Factors in the Formation of Volatile Aroma in Pears

It should be noted that the biosynthesis of esters via the LOX pathway is largely dependent on an adequate supply of substrate [49]. Aldehydes are reduced to alcohols by ADH [50], and fruit esters are then biosynthesized by AAT, with the expression of AAT being ethylene-dependent in climacteric fruits such as apple [51] and pear [6]. In our study, ADH genes (LOC103931841, LOC103965089, LOC103949386, LOC103944272, LOC103938216, LOC103940564, and novel.2114) and AAT genes (LOC103952287) in the fatty acid metabolism pathway were identified by co-expression network analysis. A large amount of ethylene is produced during the maturation and storage of Panguxiang. The AAT gene is induced by ethylene and regulates the production of volatile esters, resulting in the development of a rich fruit flavor during ripening and storage. Pears have a "green" aroma when they are not ripe, which is caused by volatile alcohol. ADH genes are the key genes in the regulation of volatile alcohol production. Our research shows that alcohol concentrations are higher when fruits are riper and stored for longer, which may result from the accumulation of ADH, providing further evidence that ADH and AAT genes play an important role in forming volatile aromas.

Supplementary Materials: The following supporting information can be downloaded at: https://www.mdpi.com/article/10.3390/life12101494/s1, Figure S1: RNA-Seq analysis of Panguxiang over different growth and development stages and storage stages. (**a**,**b**) Boxplots and violin plots of gene expression levels. (**c**) Principal component analysis and (**d**) sample correlation based on the RNA-Seq count matrix. (**e**) Heatmap of DEGs. (**f**) Numbers of differentially expressed genes in different pairs of samples. Figure S2: GO analysis diagram of growth and development period and storage period. Figure S3: KOG-enriched analysis of growth and development period and storage period. Figure S4: Analysis of KEGG pathway enrichment, growth and storage period. Figure S5: 15 co-expressed gene modules of growth and development period and storage period. Table S1: Data preprocessing results statistics. Table S2: The results of the genomic comparison. Table S3: The number of differentially expressed genes. Table S4: Summary of functional annotation. Table S5: The number of TFs in different modules. Table S6: The number and types of TFs in different modules.

Author Contributions: Conceptualization, Z.L. (Zhen Liu) and X.G.; methodology, J.Q. and Q.C.; validation, H.L.; formal analysis, H.L.; resources, Z.L. (Zhen Liu); data curation, H.L.; writing—original draft preparation, H.L.; writing—review and editing, J.Q. and S.R.; visualization, H.L. and S.R.; supervision, Y.W. and Z.L. (Zhi Li); project administration, Z.L. (Zhen Liu) and S.M.; funding acquisition, J.Q. and Z.L. (Zhen Liu). All authors have read and agreed to the published version of the manuscript.

Funding: This research was funded by the National Natural Science Foundation of China (grant number 31700549) and the Achievements Transformation Fund Project (grant number 2014D00000150).

Institutional Review Board Statement: Not applicable.

Informed Consent Statement: Not applicable.

Data Availability Statement: Not applicable.

Acknowledgments: We are grateful to the reviewers for their constructive comments and valuable suggestions, which have helped improve the quality of the paper. The authors also thank Pengcheng Li, Shunfu Li for his guidance on the production of figures in the manuscript and Huimin Wang and Lisha Fang for their advice on the experiment.

Conflicts of Interest: The authors declare no conflict of interest.

References

1. Sun, Q.; Zhang, N.; Wang, J.; Zhang, H.; Li, D.; Shi, J.; Li, R.; Weeda, S.; Zhao, B.; Ren, S.; et al. Melatonin Promotes Ripening and Improves Quality of Tomato Fruit during Postharvest Life. *J. Exp. Bot.* **2015**, *66*, 657–668. [CrossRef] [PubMed]
2. Spaho, N.; Gaši, F.; Leitner, E.; Blesić, M.; Akagić, A.; Žuljević, S.O.; Kurtović, M.; Ratković, D.Đ.; Murtić, M.S.; Akšić, M.F.; et al. Characterization of Volatile Compounds and Flavor in Spirits of Old Apple and Pear Cultivars from the Balkan Region. *Foods* **2021**, *10*, 1258. [CrossRef] [PubMed]
3. Jennings, W.G.; Sevenants, M.R. Volatile Esters of Bartlett Pear. III.a. *J. Food Sci.* **1964**, *29*, 158–163. [CrossRef]
4. Qin, G.; Wang, Y.; Cao, B.; Wang, W.; Tian, S. Unraveling the Regulatory Network of the MADS Box Transcription Factor RIN in Fruit Ripening. *Plant J.* **2012**, *70*, 243–255. [CrossRef] [PubMed]
5. Qin, G.; Tao, S.; Cao, Y.; Wu, J.; Zhang, H.; Huang, W.; Zhang, S. Evaluation of the Volatile Profile of 33 *Pyrus Ussuriensis Cultivars* by HS-SPME with GC-MS. *Food Chem.* **2012**, *134*, 2367–2382. [CrossRef] [PubMed]
6. Li, G.; Jia, H.; Li, J.; Wang, Q.; Zhang, M.; Teng, Y. Emission of Volatile Esters and Transcription of Ethylene- and Aroma-Related Genes during Ripening of 'Pingxiangli' Pear Fruit (*Pyrus Ussuriensis Maxim*). *Sci. Hortic.* **2014**, *170*, 17–23. [CrossRef]
7. Kou, X.; Liu, X.; Li, J.; Xiao, H.; Wang, J. Effects of Ripening, 1-Methylcyclopropene and Ultra-High-Pressure Pasteurisation on the Change of Volatiles in Chinese Pear Cultivars. *J. Sci. Food Agric.* **2012**, *92*, 177–183. [CrossRef]
8. Zhang, L.P.; Shen, Y.X.; Bu, Q.Z.; Ji, S.J. Effects of 1-Methylcyclopropene on the Metabolic Pathways of Aroma-Related Compounds in Nanguo Pear. *J. Food Process. Preserv.* **2014**, *38*, 1749–1758. [CrossRef]
9. Rapparini, F.; Predieri, S. Pear Fruit Volatiles. *Hortic. Rev.* **2010**, *28*, 237–324. [CrossRef]
10. Ban, Y.; Oyama-Okubo, N.; Honda, C.; Nakayama, M.; Moriguchi, T. Emitted and Endogenous Volatiles in 'Tsugaru' Apple: The Mechanism of Ester and (E,E)-α-Farnesene Accumulation. *Food Chem.* **2010**, *118*, 272–277. [CrossRef]
11. Dong, J.; Zhang, Y.; Tang, X.; Jin, W.; Han, Z. Differences in Volatile Ester Composition between *Fragaria* × *Ananassa* and *F. Vesca* and Implications for Strawberry Aroma Patterns. *Sci. Hortic.* **2013**, *150*, 47–53. [CrossRef]
12. Imahori, Y.; Yamamoto, K.; Tanaka, H.; Bai, J. Residual Effects of Low Oxygen Storage of Mature Green Fruit on Ripening Processes and Ester Biosynthesis during Ripening in Bananas. *Postharvest. Biol. Technol.* **2013**, *77*, 19–27. [CrossRef]
13. Qin, G.; Tao, S.; Zhang, H.; Huang, W.; Wu, J.; Xu, Y.; Zhang, S. Evolution of the Aroma Volatiles of Pear Fruits Supplemented with Fatty Acid Metabolic Precursors. *Molecules* **2014**, *19*, 20183–20196. [CrossRef]
14. Dudareva, N.; Klempien, A.; Muhlemann, J.K.; Kaplan, I. Biosynthesis, Function and Metabolic Engineering of Plant Volatile Organic Compounds. *New Phytol.* **2013**, *198*, 16–32. [CrossRef]
15. Defilippi, B.G.; Manríquez, D.; Luengwilai, K.; González-Agüero, M. Chapter 1 Aroma Volatiles: Biosynthesis and Mechanisms of Modulation During Fruit Ripening. *Adv. Bot. Res.* **2009**, *50*, 1–37. [CrossRef]
16. Gigot, C.; Ongena, M.; Fauconnier, M.-A.; Wathelet, J.-P.; Du Jardin, P.; Thonart, P. The Lipoxygenase Metabolic Pathway in Plants: Potential for Industrial Production of Natural Green Leaf Volatiles. 2010. Available online: https://popups.uliege.be/1780-4507/index.php?id=5669 (accessed on 20 September 2022).

17. Chen, J.L.; Yan, S.; Feng, Z.; Xiao, L.; Hu, X.S. Changes in the Volatile Compounds and Chemical and Physical Properties of Yali Pear (Pyrus Bertschneideri Reld) during Storage. *Food Chem.* **2006**, *97*, 248–255. [CrossRef]
18. Lara, I.; Miró, R.M.; Fuentes, T.; Sayez, G.; Graell, J.; López, M.L. Biosynthesis of Volatile Aroma Compounds in Pear Fruit Stored under Long-Term Controlled-Atmosphere Conditions. *Postharvest Biol. Technol.* **2003**, *29*, 29–39. [CrossRef]
19. Suwanagul, A.; Richardson, D.G. Identification of Headspace Volatile Compounds from Different Pear (*Pyrus Communis L.*) Varieties. *Acta Hortic.* **1998**, *475*, 605–623. [CrossRef]
20. Takeoka, G.R.; Buttery, R.G.; Flath, R.A. Volatile Constituents of Asian Pear (*Pyrus Serotina*). *J. Agric. Food Chem.* **1992**, *40*, 1925–1929. [CrossRef]
21. Fan, Z.; Hasing, T.; Johnson, T.S.; Garner, D.M.; Barbey, C.R.; Colquhoun, T.A.; Sims, C.A.; Resende, M.F.R.; Whitaker, V.M. Strawberry Sweetness and Consumer Preference Are Enhanced by Specific Volatile Compounds. *Hortic. Res.* **2021**, *8*, 1–15. [CrossRef]
22. Chang, S.; Puryear, J.; Cairney, J. A Simple and Efficient Method for Isolating RNA from Pine Trees. *Plant Mol. Biol. Report.* **1993**, *11*, 113–116. [CrossRef]
23. Chen, S.; Zhou, Y.; Chen, Y.; Gu, J. Fastp: An Ultra-Fast All-in-One FASTQ Preprocessor. *Bioinformatics* **2018**, *34*, i884–i890. [CrossRef]
24. Kim, D.; Langmead, B.; Salzberg, S.L. HISAT: A Fast Spliced Aligner with Low Memory Requirements. *Nat. Methods* **2015**, *12*, 357–360. [CrossRef]
25. Liao, Y.; Smyth, G.K.; Shi, W. FeatureCounts: An Efficient General Purpose Program for Assigning Sequence Reads to Genomic Features. *Bioinformatics* **2014**, *30*, 923–930. [CrossRef]
26. Varet, H.; Brillet-Guéguen, L.; Coppée, J.Y.; Dillies, M.A. SARTools: A DESeq2- and EdgeR-Based R Pipeline for Comprehensive Differential Analysis of RNA-Seq Data. *PLoS ONE* **2016**, *11*, e0157022. [CrossRef]
27. Ali, M.; Hussain, R.M.; Rehman, N.U.; She, G.; Li, P.; Wan, X.; Guo, L.; Zhao, J.; Nishimura, M. De Novo Transcriptome Sequencing and Metabolite Profiling Analyses Reveal the Complex Metabolic Genes Involved in the Terpenoid Biosynthesis in Blue Anise Sage (*Salvia guaranitica* L.). *DNA Res.* **2018**, *25*, 597–617. [CrossRef]
28. Apweiler, R.; Bairoch, A.; Wu, C.H.; Barker, W.C.; Boeckmann, B.; Ferro, S.; Gasteiger, E.; Huang, H.; Lopez, R.; Magrane, M.; et al. UniProt: The Universal Protein Knowledgebase. *Nucleic Acids Res.* **2004**, *32*, D115–D119. [CrossRef]
29. Bateman, A.; Martin, M.J.; O'Donovan, C.; Magrane, M.; Alpi, E.; Antunes, R.; Bely, B.; Bingley, M.; Bonilla, C.; Britto, R.; et al. UniProt: The Universal Protein Knowledgebase. *Nucleic Acids Res.* **2017**, *45*, D158–D169. [CrossRef]
30. Ashburner, M.; Ball, C.A.; Blake, J.A.; Botstein, D.; Butler, H.; Cherry, J.M.; Davis, A.P.; Dolinski, K.; Dwight, S.S.; Eppig, J.T.; et al. Gene Ontology: Tool for the Unification of Biology. *Nat. Genet.* **2000**, *25*, 25–29. [CrossRef]
31. Koonin, E.V.; Fedorova, N.D.; Jackson, J.D.; Jacobs, A.R.; Krylov, D.M.; Makarova, K.S.; Mazumder, R.; Mekhedov, S.L.; Nikolskaya, A.N.; Rao, B.S.; et al. A Comprehensive Evolutionary Classification of Proteins Encoded in Complete Eukaryotic Genomes. *Genome Biol.* **2004**, *5*, 1–28. [CrossRef]
32. Finn, R.D.; Bateman, A.; Clements, J.; Coggill, P.; Eberhardt, R.Y.; Eddy, S.R.; Heger, A.; Hetherington, K.; Holm, L.; Mistry, J.; et al. Pfam: The Protein Families Database. *Nucleic Acids Res.* **2014**, *42*, D222–D230. [CrossRef] [PubMed]
33. Kanehisa, M.; Goto, S.; Kawashima, S.; Okuno, Y.; Hattori, M. The KEGG Resource for Deciphering the Genome. *Nucleic Acids Res.* **2004**, *32*, D277–D280. [CrossRef] [PubMed]
34. O'Donovan, C.; Martin, M.J.; Glemet, E.; Codani, J.J.; Apweiler, R. Removing Redundancy in SWISS-PROT and TrEMBL. *Bioinformatics* **1999**, *15*, 258–259. [CrossRef] [PubMed]
35. Bairoch, A.; Apweiler, R. The SWISS-PROT Protein Sequence Database and Its Supplement TrEMBL in 2000. *Nucleic Acids Res.* **2000**, *28*, 45–48. [CrossRef] [PubMed]
36. Zheng, Y.; Jiao, C.; Sun, H.; Rosli, H.G.; Pombo, M.A.; Zhang, P.; Banf, M.; Dai, X.; Martin, G.B.; Giovannoni, J.J.; et al. ITAK: A Program for Genome-Wide Prediction and Classification of Plant Transcription Factors, Transcriptional Regulators, and Protein Kinases. *Mol. Plant* **2016**, *9*, 1667–1670. [CrossRef] [PubMed]
37. Langfelder, P.; Horvath, S. WGCNA: An R Package for Weighted Correlation Network Analysis. *BMC Bioinform.* **2008**, *9*, 1–13. [CrossRef]
38. Shannon, P.; Markiel, A.; Ozier, O.; Baliga, N.S.; Wang, J.T.; Ramage, D.; Amin, N.; Schwikowski, B.; Ideker, T. Cytoscape: A Software Environment for Integrated Models of Biomolecular Interaction Networks. *Genome Res.* **2003**, *13*, 2498–2504. [CrossRef]
39. Livak, K.J.; Schmittgen, T.D. Analysis of Relative Gene Expression Data Using Real-Time Quantitative PCR and the 2−ΔΔCT Method. *Methods* **2001**, *25*, 402–408. [CrossRef]
40. Bustin, S.A.; Benes, V.; Garson, J.A.; Hellemans, J.; Huggett, J.; Kubista, M.; Mueller, R.; Nolan, T.; Pfaffl, M.W.; Shipley, G.L.; et al. The MIQE Guidelines: Minimum Information for Publication of Quantitative Real-Time PCR Experiments. *Clin. Chem.* **2009**, *55*, 611–622. [CrossRef]
41. Abdullah-Zawawi, M.R.; Ahmad-Nizammuddin, N.F.; Govender, N.; Harun, S.; Mohd-Assaad, N.; Mohamed-Hussein, Z.A. Comparative Genome-Wide Analysis of WRKY, MADS-Box and MYB Transcription Factor Families in Arabidopsis and Rice. *Sci. Rep.* **2021**, *11*, 1–18. [CrossRef]
42. Saltveit, M.E. Effect of Ethylene on Quality of Fresh Fruits and Vegetables. *Postharvest Biol. Technol.* **1999**, *15*, 279–292. [CrossRef]
43. Li, G.; Jia, H.; Wu, R.; Hussain, S.; Teng, Y. Characterization of Aromatic Volatile Constituents in 11 Asian Pear Cultivars Belonging to Different Species. *Afr. J. Agric. Res.* **2012**, *7*, 4761–4770. [CrossRef]

44. López-Nicolás, J.M.; Andreu-Sevilla, A.J.; Carbonell-Barrachina, Á.A.; García-Carmona, F. Effects of Addition of α-Cyclodextrin on the Sensory Quality, Volatile Compounds, and Color Parameters of Fresh Pear Juice. *J. Agric. Food Chem.* **2009**, *57*, 9668–9675. [CrossRef]
45. Moya-León, M.A.; Vergara, M.; Bravo, C.; Montes, M.E.; Moggia, C. 1-MCP Treatment Preserves Aroma Quality of 'Packham's Triumph' Pears during Long-Term Storage. *Postharvest Biol. Technol.* **2006**, *42*, 185–197. [CrossRef]
46. Yang, Y.; Han, L.; Yuan, Y.; Li, J.; Hei, N.; Liang, H. Gene Co-Expression Network Analysis Reveals Common System-Level Properties of Prognostic Genes across Cancer Types. *Nat. Commun.* **2014**, *5*, 1–9. [CrossRef]
47. El-Sharkawy, I.; Liang, D.; Xu, K. Transcriptome Analysis of an Apple (*Malus* × *Domestica*) Yellow Fruit Somatic Mutation Identifies a Gene Network Module Highly Associated with Anthocyanin and Epigenetic Regulation. *J. Exp. Bot.* **2015**, *66*, 7359–7376. [CrossRef]
48. Yuan, L.; Chen, L.; Qian, K.; Qian, G.; Wu, C.L.; Wang, X.; Xiao, Y. Co-Expression Network Analysis Identified Six Hub Genes in Association with Progression and Prognosis in Human Clear Cell Renal Cell Carcinoma (CcRCC). *Genom. Data* **2017**, *14*, 132–140. [CrossRef]
49. Song, J.; Bangerth, F. Fatty Acids as Precursors for Aroma Volatile Biosynthesis in Pre-Climacteric and Climacteric Apple Fruit. *Postharvest Biol. Technol.* **2003**, *30*, 113–121. [CrossRef]
50. Echeverría, G.; Fuentes, T.; Graell, J.; Lara, I.; López, M.L. Aroma Volatile Compounds of 'Fuji' Apples in Relation to Harvest Date and Cold Storage Technology: A Comparison of Two Seasons. *Postharvest Biol. Technol.* **2004**, *32*, 29–44. [CrossRef]
51. Schaffer, R.J.; Friel, E.N.; Souleyre, E.J.F.; Bolitho, K.; Thodey, K.; Ledger, S.; Bowen, J.H.; Ma, J.H.; Nain, B.; Cohen, D.; et al. A Genomics Approach Reveals That Aroma Production in Apple Is Controlled by Ethylene Predominantly at the Final Step in Each Biosynthetic Pathway. *Plant Physiol.* **2007**, *144*, 1899–1912. [CrossRef]

life

MDPI

Article

Molecular Cloning and Functional Analysis of *IrUGT86A1-like* Gene in Medicinal Plant *Isodon rubescens* (Hemsl.) Hara

Conglong Lian, Jinxu Lan, Bao Zhang, Hao Yang, Kaihua Guo, Jingjing Li and Suiqing Chen *

School of Pharmacy, Henan Key Laboratory of Chinese Medicine Resources and Chinese Medicine Chemistry, Henan University of Chinese Medicine, Boxue Road, Jinshui District, Zhengzhou 450046, China
* Correspondence: chsq@hactcm.edu.cn

Abstract: The synthesis of secondary metabolites in plants often includes glycosylation modifications. Often, the final step of constructing plant secondary metabolites is completed by glycosylation transferases, which are also involved in many cell processes. In this study, a UDP-glycosyltransferase gene (*UGT*) was amplified from *Isodon rubescens* (Hemsl.) Hara with RT-PCR and named *IrUGT86A1-like* (GenBank: MZ913258). Here, we found that *IrUGT86A1-like* gene is 1450 bp in length and encodes for 479 amino acids. Bioinformatics analysis revealed that *IrUGT86A1-like* is a stable and hydrophilic protein, located in the cytoplasm with a transmembrane domain. Phylogenetic analysis showed that IrUGT86A1-like protein has the closest genetic relationship with the UDP-glycosyltransferase 86A1-like protein (XP_042054241.1) of *Salvia splendens*. RT-qPCR analysis demonstrated that the expression of *IrUGT86A1-like* gene varied in different tissues; leaves had the highest expression followed by flowers, stems, and roots had the lowest expression. This expression trend is similar to the distribution of oridonin content in different tissues of *I. rubescens*. Additionally, *IrUGT86A1-like* gene was found to be positively enhanced by NaCl and MeJA treatment, and in contrast was down-regulated by ABA treatment. Finally, the prokaryotic expression vector pEASY®-Blunt E1-IrUGT86A1 was successfully used to express about 53 KD of IrUGT86A1-like protein. This research builds a foundation for further investigation on the function of this gene in the synthesis and modification of secondary metabolites.

Keywords: *Isodon rubescens*; UDP-glycosyltransferase; RT-qPCR; prokaryotic expression

Citation: Lian, C.; Lan, J.; Zhang, B.; Yang, H.; Guo, K.; Li, J.; Chen, S. Molecular Cloning and Functional Analysis of *IrUGT86A1-like* Gene in Medicinal Plant *Isodon rubescens* (Hemsl.) Hara. *Life* **2022**, *12*, 1334. https://doi.org/10.3390/life12091334

Academic Editor: Stefania Lamponi

Received: 29 July 2022
Accepted: 25 August 2022
Published: 28 August 2022

1. Introduction

The "Donglingcao", a Chinese herb, refers to the dry aboveground part of the Labiatae plant *Isodon rubescens* (Hemsl.) Hara, and is a commonly used for its medicinal properties. Harvested in summer and autumn when the stems and leaves are lush, "Donglingcao" tastes both bitter and sweet, and is consumed slightly cold [1]. It is used for inflammation, antipyretic detoxification, promoting blood circulation, promoting stomach health, and anti-tumor effects. The herb is commonly used in the treatment of sore throat, bronchitis, acute laryngitis, acute suppurative tonsillitis, abdominal mass, and snake bite, [1,2]. Modern pharmacological studies have shown that the main active ingredients in *I. rubescens* plants are diterpene components, oridonin, ponicidin, and water-soluble components, such as rosmarinic acid, among which oridonin is the most active component in the pharmacopoeia [3]. These active ingredients of *I. rubescens* confer significant curative effects on breast cancer, esophageal cancer, colon cancer, and other cancers [4]. Therefore, as an anti-tumor plant with evidence of strong pharmacological activity and low hepatorenal toxicity, *I. rubescens* has been a research focus of pharmacological scientists worldwide regarding the development of molecular pharmacognosy technology and the analysis of the biosynthesis of the main components of *I. rubescens*, including oridonin [5]. In addition to *I. rubescens*' medicinal properties, it is also a strong and hardy plant with characteristics such as drought resistance, cold resistance, barren resistance, developed root system, and soil amendments and water conservation effects [6].

A form of translational and post-translational modification, glycosylation is an enzyme-mediated process in which sugars are added to proteins or lipids [7]. Glycosylation is not only a key factor in maintaining the stability of the intracellular environment; it also plays an important role in the diversity and complexity of plant secondary metabolites synthesis. Specifically, glycosyltransferases (GTs) are the main enzymes in glycosylation [8,9]. In living organisms, catalyzed activated sugars are connected to different receptor molecules, which contributes to the hypothesis that obtained glycosylation products have many biological functions, and thus regulate various biological functions of plants [9,10]. Among them, GT1 plants are the largest of the GT family. As GT1s use uridine diphosphate-glycosyltransferases as a donor molecule, GT1s are also known as "UGTs" [11]. Previous studies in plants have shown that UGTs are involved in the biosynthesis of terpenoids, phenols, flavonoids, steroids, and other natural products [12]. Furthermore, UGTs catalyze the glycosyltransferase-mediated conversion of uridine diphosphate-activated sugars to substrates such as hormones and secondary metabolites, thus contributing to diverse and complex substrates in plants, which in turn regulate various biological processes such as plant growth, development, disease resistance, and abiotic stress resistance [13–15]. In *Arabidopsis*, *UGT79B1* and *UGT91A1* are specifically involved in the modification processing steps of anthocyanin synthesis, and they also contribute to cold, salt, and drought stress tolerance [16]. Similarly, in wheat, overexpression of *Ta-UGT3* gene significantly enhances *Fusarium Head* Blight resistance [17]. Moreover, in rice, a UDP-glucosyltransferase gene regulates rice grain size and biological stress response by changing the reflow of plant metabolic flow [18]. Additionally, *UGT85C2*, *UGT74G1*, and *UGT76G1* in *Stevia rebaudiana*, have been found to have activity in the synthesis of stevioside and rebaudioside A, which are two main glycosides of *Stevia* [19]. Furthermore, a UDP-Glucosyltransferase gene was found to be beneficial to the biosynthesis of flavonoid glycosides in *Cyclocarya paliurus* [20].

Because of the high abundance of terpenoids, flavonoids, alkaloids, steroids, volatile oils, amino acids, organic acids, monosaccharides, and other components in *I. rubescens* [21], glycosylation plays an important role in these plants as it is essential to the synthesis and modification of these compounds. However, research on glycosyltransferase in *I. rubescens* has not yet been reported. Therefore, it is essential to investigate the glycosyltransferase of *I. rubescens* in order to understand the biosynthesis of important secondary metabolites. In a previous transcriptome data analysis with two tissues: callus with undetectably low levels oridonin and adventitious buds with oridonin, a significant difference in the expression in UDP-Glucosyltransferase (*UGT*) gene was found, suggesting that it may modify and regulate oridonin synthesis. With these previous transcriptomic analysis results, this significant differential expression in the *UGT* sequence was used as a reference, and this *UGT* candidate gene was cloned by RT-PCR. The physicochemical properties, protein structure, and evolutionary relationship were analyzed by the bioinformatics. The differential expression in different tissues and responses to different abiotic stresses were further analyzed with fluorescent quantitative PCR (RT-qPCR). Finally, the protein was successfully expressed in *E. coli* bacteria. Overall, this foundational research enables further investigation into the role of the enzyme system encoded by *IrUGT86A1-like* gene in the synthesis of active ingredients in *I. rubescens* and research on the function of this gene with genetic engineering.

2. Materials and Methods

2.1. Plant Material and Sample Preparation

I. rubescens plants were cultivated in artificial intelligence climate boxes with the following conditions: nutrient soil and vermiculite (2:1), photoperiod 14 h/10 h (light/dark), relative humidity 60%, and culture temperature 25 °C. After six months of growth at flowering, the roots, stems, leaves, and flowers of *I. rubescens* were collected for quantification of oridonin and RNA extraction (stored at −80 °C). For the simulated salt stress treatment, uniformly developed seedlings of *I. rubescens* were fully watered using a 200 mM NaCl solution, and the leaves between three and six nodes were harvested at 0, 2, 8, 12, 24, and

72 h. For the MeJA and ABA treatments, 200 μmol MeJA, and 300 μmol ABA, were evenly sprayed on the leaves of *I. rubescens*, respectively, and the leaves between three and six nodes were harvested at six time points at 0, 2, 4, 8, 12, and 24 h. All collected samples were frozen in liquid nitrogen immediately and then preserved at −80 °C for RNA extraction and expression analysis of the *IrUGT86A1-like* gene.

2.2. Total RNA Extraction and cDNA Reverse Transcription

Frozen samples of roots, stems, leaves, flowers, and other tissues of *I. rubescens* were individually ground into a fine powder with a pestle and mortar using liquid nitrogen. The total RNA was extracted according to the EASYspin Plus plant RNA extraction kit protocol (AidLab, Beijing, China). An ultramicro spectrophotometer was used to measure quality and quantity of RNA, and all RNA samples were stored at −80 °C refrigerator for downstream analysis. Next, the reverse transcription kit BCS HIScript^{TM} AII-in-One Mix with dsDNAse was to remove genomic DNA contamination, and then to synthesize the first strand of cDNA was synthesized via reverse transcription. The obtained cDNA was adjusted to the corresponding concentration according to quantification with the ultramicro spectrophotometer and stored in a refrigerator at −20 °C for further use.

2.3. Cloning of IrUGT86A1-like Gene

First, primers of UGT-F and UGT-R were designed, and the first strand of *I. rubescens* cDNA was used in cloning to obtain *IrUGT86A1-like* gene sequence. The PCR reaction consisted of 2× Taq Plus Master Mix 12.5 μL, cDNA 1 μL, UGT-F 1 μL, UGT-R 1 μL, and ddH_2O 9.5 μL. The PCR procedure was as follows: pre-denaturation at 95 °C for 3 min; denaturation at 95 °C for 15 s, annealing at 53 °C for 30 s, extension at 72 °C for 90 s, 35 cycles; and extension at 72 °C for 10 min. The PCR products were detected with 1% agarose gel electrophoresis in which DNA was separated based on length, and target DNA band was recovered with SanPrep Column DNA Gel Extraction Kit (Sangon Biotech, Shanghai, China). The obtained target fragment was further ligated using the cloning vector pEASY®-Blunt Zero Cloning Kit (Transgen, Beijing, China), and transformed into *E. coli* T1 competent cells. The positive clones were selected and sent to Sangon Biotech (Shanghai) Co., Ltd. for sequencing, and the correct pEASY-Blunt-IrUGT86A1 prokaryotic strain was obtained. The primers for PCR are included in Table S1.

2.4. Bioinformatics Analysis

DNAMAN6.0 software was used to analyze sequencing results and investigate and predict proteins. Conserved domain prediction was performed with the NCBI online tool Conserved Domains (https://www.ncbi.nlm.nih.gov/cdd, accessed on 6 May 2022). Protein physical and chemical properties were predicted using the online tool ExPASy (http://www.expasy.ch/tools/protparam.html, accessed on 6 May 2022). Subcellular localization was predicted and analyzed using the online tools WoLF PSORT (https://wolfpsort.hgc.jp/, accessed on 6 May 2022) and PredictProtein (https://predictprotein.org/, accessed on 6 May 2022) results. Protein signal peptides were predicted using online tools SignalP 5.0 Server (http://www.cbs.dtu.dk/services/SignalP/, accessed on 7 May 2022) and TargetP-2.0 (http://www.cbs.dtu.dk/services/TargetP/, accessed on 7 May 2022). The protein transmembrane domain was predicted with TMPred (https://www.expasy.org/resources/tmpred, accessed on 7 May 2022) and the TMHMM Server v.2.0 (http://www.cbs.dtu.dk/services/TMHMM/, accessed on 7 May 2022). Protein secondary structure and coil structure were analyzed using the online tools PredictProtein (https://predictedprotein.org/, accessed on 8 May 2022) and COILS (https://embnet.vital-it.ch/software/COILS_form.html, accessed on 8 May 2022). Three-dimensional modeling of protein domains was performed using SWISS-MODEL (http://swissmodel.expasy.org/, accessed on 8 May 2022).

2.5. Multi-Sequence Alignment and Construction of Phylogenetic Tree

Similar amino acid sequences to the *IrUGT86A1-like* gene encoding protein of *I. rubescens* were obtained using blastp on the NCBI database, and the amino acid sequences of UGT86A1 from other plants were downloaded from NCBI database. These UGT86A1 protein sequences including IrUGT86A1-like of *I. rubescens* were processed with to multiple sequence alignment analysis using MEGA7.0 software [22], and the phylogenetic tree was constructed using the neighbor-joining method with the following parameters: bootstrap = 1000 repeat times, and default values were used for other parameters.

2.6. Fluorescence Quantitative Expression Analysis

The cDNA of roots, stems, leaves, flowers, and callus of *I. rubescens* were used as templates, while *GADPH* was used as the internal reference gene for RT-qPCR of the *IrUGT86A1-like* gene. The reaction system was as follows: 10 μL of 2 × TransStart® Top Green qPCR SuperMix, 0.4 μL of 50 × Passive Reference Dye II, and 0.4 μL of forward and reverse primers (10 μM), respectively, sample cDNA 2 μL, and ddH_2O 6.8 μL. Reaction conditions were the following: 94 °C 30 s; 94 °C 5 s, 60 °C 30 s, and 45 cycles. qPCR was repeated three times, and the relative expression of genes was calculated by $2^{-\Delta\Delta Ct}$ method [23]. The primers for RT-qPCR are included in Table S1.

2.7. Determination of Oridonin in I. rubescens

Oridonin was extracted and detected according to the reference: Pharmacopoeia of the People's Republic of China [1]. The roots, stems, leaves, and flowers of *I. rubescens* plants were respectively placed in an oven at 60 °C and baked to a constant weight. Then they were powdered with a mortar. We accurately weighed 1 g of this product powder, put it in a conical flask with a stopper, accurately added 50 mL methanol, weighed and set the weight, placed it for 30 min, gave it ultrasonic treatment (power 250 W, frequency 40 kHz) for 30 min, made up the lost weight with methanol, shook it well, and filtered 0.22 μm to obtain the sample to be tested. Preparation of control solution: an appropriate amount of isodon reference solution was taken, accurately weighed, and methanol was added to make a solution containing 60 μg per 1 mL. An amount of 10 μL of each of the control solution and the test solution were precisely absorbed and injected into the liquid chromatograph for determination. The chromatographic conditions were the following: Waters e2695/2998, Phenomenex C18 column (250 mm × 4.6 mm, 4 um), mobile phase methanol–water (55:45), detection wavelength 239 nm, flow rate 0.8 mL·min^{-1}, and column temperature 25 °C.

2.8. Prokaryotic Expression Analysis

Using the sequence of the *IrUGT86A1-like* gene and characteristics of the pEASY®-Blunt E1 Expression Vector, a pair of prokaryotic expression primers UGT-YF and UGT-YR for *IrUGT86A1-like* gene were designed with DNAMAN software. The ORF sequence of *IrUGT86A1-like* gene was amplified with the TransStart® FastPfu Fly DNA Polymerase High Fidelity DNA Polymerase Kit (TransGen, Beijing, China) and cloned into the prokaryotic expression vector pEASY®-Blunt E1 Expression Vector (TransGen, Beijing, China). The recombinant prokaryotic expression vector pEASY-Blunt E1-IrUGT86A1 with 6×HIS-IrUGT86A1-like fusion protein was obtained with PCR identification, sequencing identification, and transformed into BL21 (DE3) prokaryotic expression competent cells. Protein expression was induced by 0.2 mmol/L IPTG for 16 h in the constant temperature vibration incubator with 200 rpm at 18 °C, and SDS-PAGE protein electrophoresis was performed to analyze the expression results.

3. Results

3.1. Cloning of IrUGT86A1-like Gene from I. rubescens

A *UGT* candidate gene sequence was obtained with bioinformatics analysis of second-generation transcriptome data. Primers were designed and the cDNA from *I. rubescens* leaves was used as the template for cloning. RT-PCR gel electrophoresis results confirmed that the target fragment of about 1500 bp was amplified (Figure 1). Sequencing results showed that the obtained band was 1450 bp in length, containing a 10 bp 5′ untranslated region (5′ UTR) and a 1440 bp open reading frame (ORF), which encodes for 479 amino acids. The blastp comparison results from NCBI showed that the cloned *IrUGT86A1-like* gene had high homology (query coverage greater than 95%) with UDP-glycosyltransferase 86A1-like genes in the database, indicating that the successfully cloned gene is a *UGT* gene of *I. rubescens* and accordingly was named *IrUGT86A1-like*. Sequences are available on NCBI (Genbank Accession Number: MZ913258).

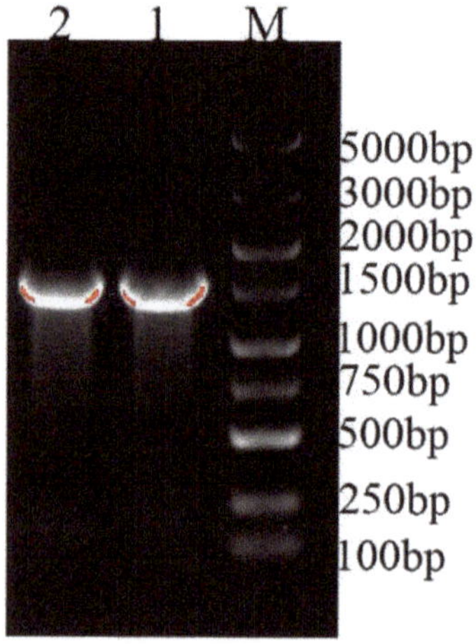

Figure 1. PCR amplification result of *IrUGT86A1-like* gene. M: 5000 ladder marker; 1–2: Amplification bands of *IrUGT86A1-like*.

3.2. Physicochemical Properties and Localization Analysis of IrUGT86A1-like Protein

ExPASy analysis showed that IrUGT86A1-like protein encodes 479 amino acids, has a molecular weight of 53.03 kD, theoretical isoelectric point of 5.33, molecular formula of $C_{2404}H_{3723}N_{633}O_{708}S_{15}$, and contains 7483 atoms. There were 59 negatively charged residues (Asp + Glu) and 41 positively charged residues (Arg + Lys). The instability coefficient is 36.16, the fat solubility coefficient is 94.82, and half-life length in *E. coli* is >10 h. The total average hydrophilicity is −0.057 indicating that the protein encoded by IrUGT86A1-like protein is a stable hydrophilic protein. Signal peptide analysis of SignalP-5.0 and TargetP-2.0 did not detect a signal peptide on this protein. WoLF PSORT protein subcellular localization prediction revealed that there are eight localization sites in the cytoplasm, four sites in the chloroplast, and two sites in the nucleus. Further analysis of protein subcellular localization with PredictProtein predicted that the protein is also located in the cytoplasm (Table S2).

The program TMpred was further used to predict the transmembrane region of IrUGT86A1-like protein. The results showed that there are seven transmembrane regions from inside to outside, and six transmembrane regions from outside to inside. This predicted transmembrane topological model contained four strong transmembrane helix regions at the N-terminal, namely, the inner–outer transmembrane helix region at 23~41 aa, the outer–inner transmembrane helix region at 140~158 aa, the inner–outer transmembrane helix region at 339~364 aa, and the outer–inner transmembrane helix region at 364~387 aa (Figure 2A). The prediction results with the TMHMM Server v. 2.0 revealed that there is a transmembrane region of N-terminal in the cytoplasm (Figure 2B).

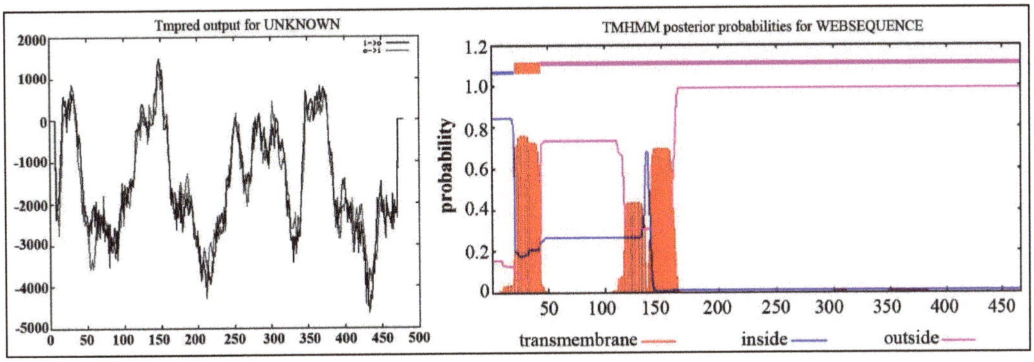

Figure 2. Transmembrane structure prediction of IrUGT86A1-like protein by TMpred (**A**) and TMHMM Server (**B**).

3.3. Structural Analysis of IrUGT86A1-like Protein from I. rubescens

Conserved domains analysis in NCBI (Figure 3A) demonstrated that the protein had GT1_Gtf-like site-specific and non-specific sites of PLN02448 and UDPGT, which indicted that IrUGT86A1-like belongs to the Glycosyltransferase_GTB-type superfamily. The predicted secondary structure of the IrUGT86A1-like protein according to PredictProtein is shown in Figure 3B, in which 33.82% of the secondary structure is composed of and alpha-helix structure (H, α-helix), 11.27% is strand structure (E, β-sheet), and the remaining 54.91% is categorized as "other structure". Solvent-accessibility predicted that 38.41% of the protein is exposed, 54.91% is buried, and 6.68% is intermediate. Therefore, these results predicted that IrUGT86A1-like protein is a mixed protein. The COILS prediction indicates that there are seven coiled helix regions when window = 14. Moreover, when window = 21, there are three coiled regions. In the case of window = 28, there is a coil region (Figure 3C). Finally, SWISS-MODEL was used to predict the three-dimensional structure of IrUGT86A1-like protein. As shown in Figure 3D, the UGT85H2 (ID: 2pq6.1) of *Medicago truncatula* was used as the template for system modeling, and the modeling range of IrUGT86A1-like protein was 7~465 aa. The model coverage was 95%, sequence similarity was 37%, and identity coincidence was 76.35%. It is evident in the tertiary structure diagram that the structure composition is consistent with the secondary structure, mainly composed of alpha-helix and loop/coil.

3.4. Phylogenetic Analysis of IrUGT86A1-like in I. rubescens

The amino acid sequence of IrUGT86A1-like protein was uploaded and searched against the nr non-redundant protein database with using the online software Protein BLAST from NCBI, and other plant UGT86A1 proteins with high homology to the query amino acid sequence of IrUGT86A1-like in *I. rubescens* were obtained. The amino acid sequences of IrUGT86A1-like from *I. rubescens*, *Salvia splendens*, *Sesamum indicum*, *Phthemospermum japonicum*, *Scoparia dulcis*, *Strobilanthes cusia*, *Morella rubra*, *Camellia sinensis*, *Morus notabilis*, *Ziziphus jujuba*, and *Hevea brasiliensis* were compared with MEGA software. The UGT86A1 protein of different species was found to be highly conserved in the conserved region (Figure 4).

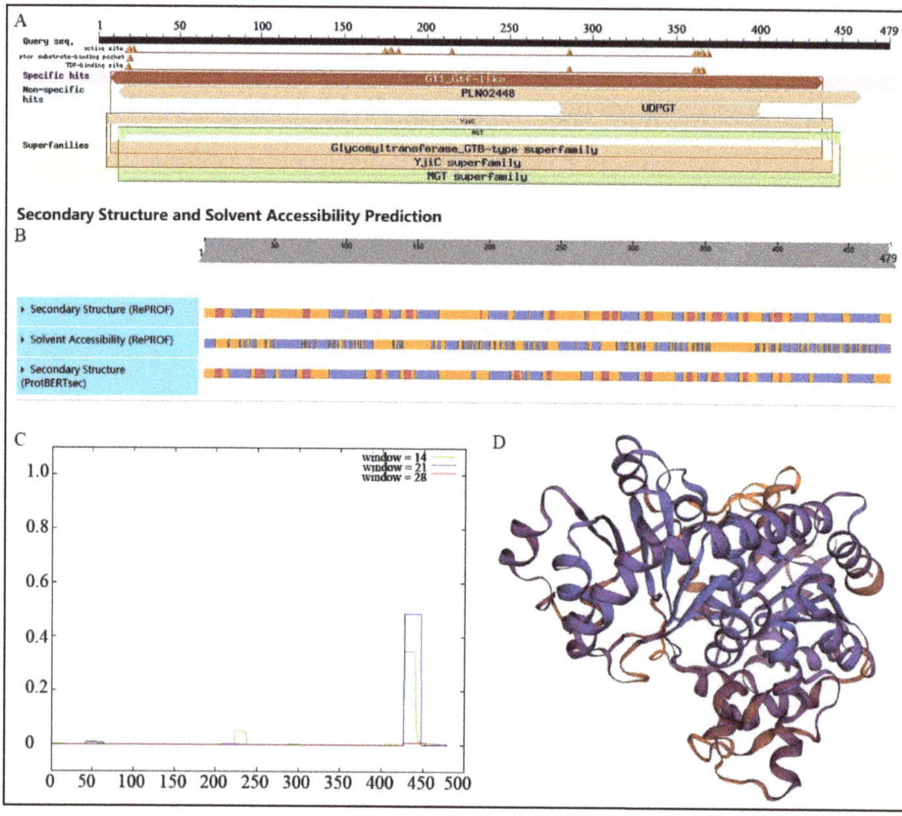

Figure 3. Structural analysis of IrUGT86A1-like protein. Conserved domains (**A**), secondary structure (**B**), COILS analysis (**C**), and 3 D structure (**D**) prediction of IrUGT86A1-like protein.

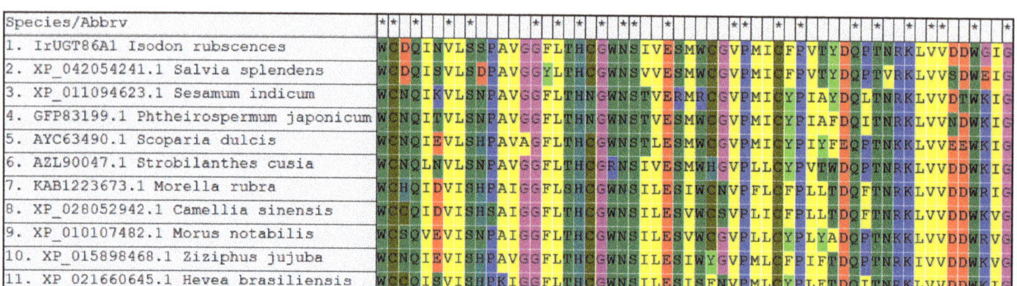

Figure 4. Conserved regions alignment analysis of IrUGT86A1-like protein homologous sequences.

MEGA7 was further used to construct the phylogenetic tree of the aforementioned 11 UGT86A1 protein sequences, as shown in Figure 5. Phylogenetic analysis revealed that the 11 UGT86A1 protein sequences were generally divided into two branches: woody plants and herbaceous plants. The IrUGT86A1-like protein of *I. rubescens* belongs to the herbaceous branch, grouping with the Labiatae and most closely genetically similar to *Salvia splendens*. Furthermore, the IrUGT86A1-like protein of *I. rubescens* has close genetic relationships with *Strobilanthes cusia*, *Phthemospermum japonicum*, and *Sesamum indicum*. *Scoparia dulcis* is an erect herb or semi-shrub-like herb, also belonging to this branch of herbs.

As a transitional stage between herbs and woody plants, *Scoparia dulcis* is closer to woody plants than *I. rubescens*. The evolutionary relationships of the IrUGT86A1-like protein found here are reflective of the evolutionary relationships between herbs and woody plants.

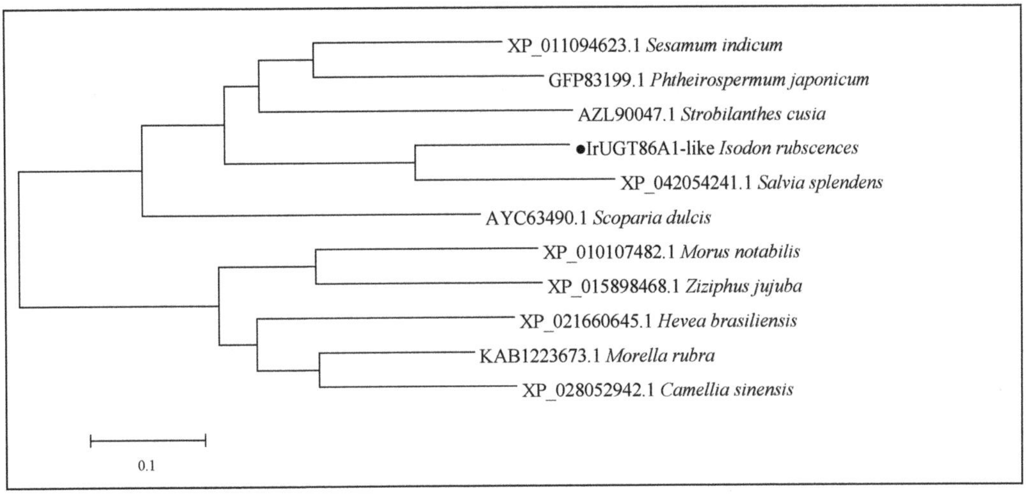

Figure 5. Phylogentic tree construction of IrUGT86A1-like protein. The black dot represents the target IrUGT86A1-like protein.

3.5. *IrUGT86A1-like Expression and Oridonin Content in Different Tissues of I. rubescens*

The expression of *IrUGT86A1-like* gene in different tissues of *I. rubescens* was detected and measured with RT-qPCR (Figure 6A). Leaves had the highest expression levels of *IrUGT86A1-like* gene, about 9.95-fold than that in roots. Flowers had the second-highest expression levels, about 5.33-fold than that in roots. Stems expressed about 1.84-fold as much as roots.

In addition to the expression profiles of the *IrUGT86A1-like* gene in different tissues of *I. rubescens*, the content of oridonin in these tissues was determined. Leaves contained the greatest oridonin content, followed by flowers, then stems, and roots had the extremely low oridonin content (Figure 6B,C). These results showed that the content of oridonin in different tissues varied with the expression level of the *IrUGT86A1-like* gene in different tissues.

3.6. *IrUGT86A1-Like Expression under Different Abiotic Treatments of I. rubescens*

Expression of the *UGT* gene has been reported increase in plants subjected to various stress conditions. Herein, the expression of *IrUGT86A1-like* under abiotic stress was further investigated here, and our results demonstrated that *IrUGT86A1-like* expression changed with NaCl, ABA, or MeJA treatments (Figure 7). For NaCl and MeJA treatments, the expression of *IrUGT86A1-like* gene initially increased and then decreased over time. The expression of *IrUGT86A1-like* gene was the greatest after 12 h of NaCl treatment, which was 3.03-fold higher than the baseline expression of the untreated control (Figure 7A). The expression of *IrUGT86A1-like* was the highest after 8 h of MeJA treatment, which was 2.37-fold higher than the untreated control expression level (Figure 7B). In regards to the ABA treatment, the expression of *IrUGT86A1-like* gene was significantly decreased, and the expression was the lowest 2 h after ABA spraying, indicating that *IrUGT86A1-like* gene is sensitive to ABA (Figure 7C).

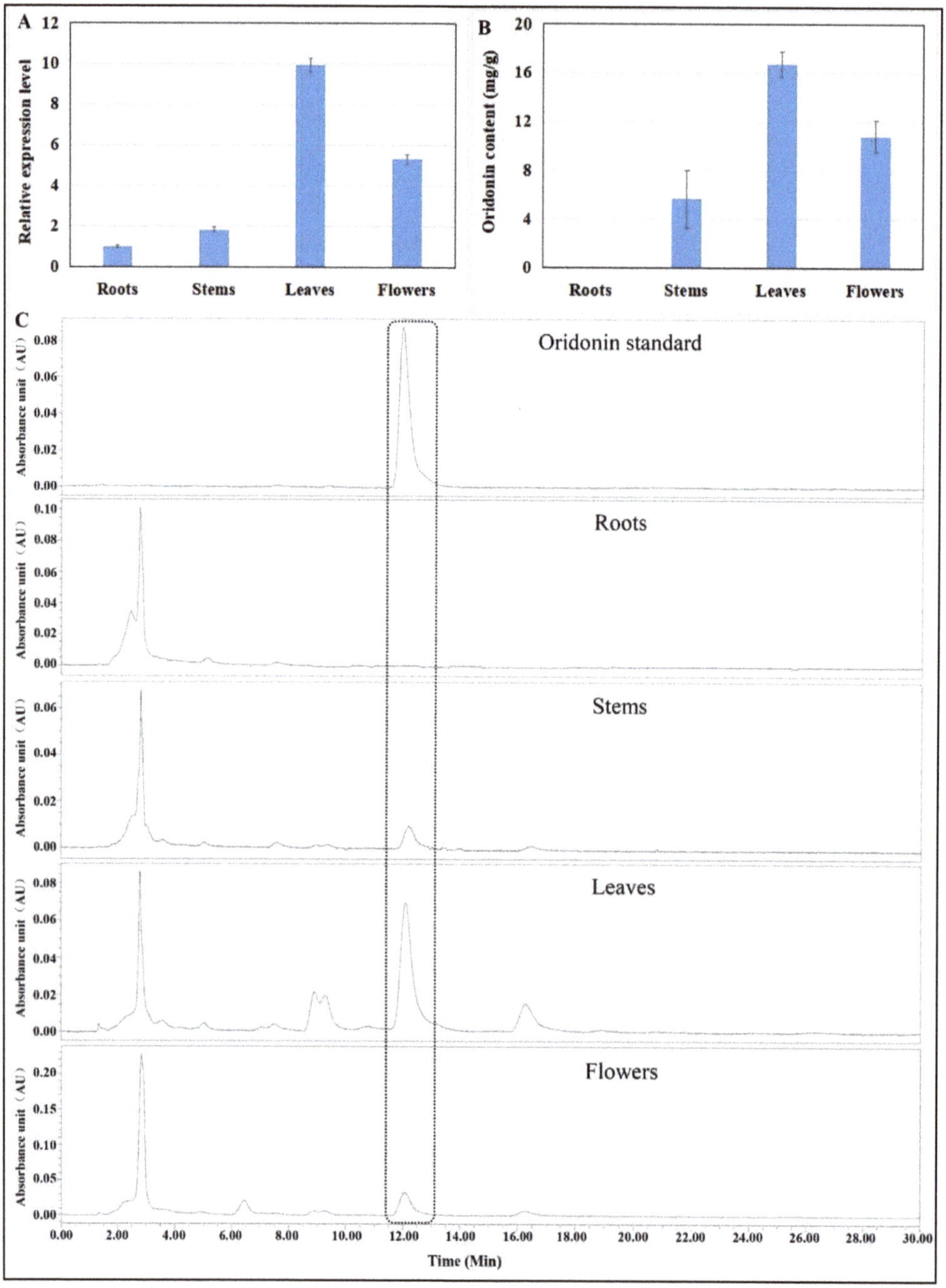

Figure 6. *IrUGT86A1-like* gene expression (**A**), oridonin content (**B**), and HPLC chromatogram (**C**) in different tissues of *I. rubescens*. Results presented by mean value of triplicate ± SE.

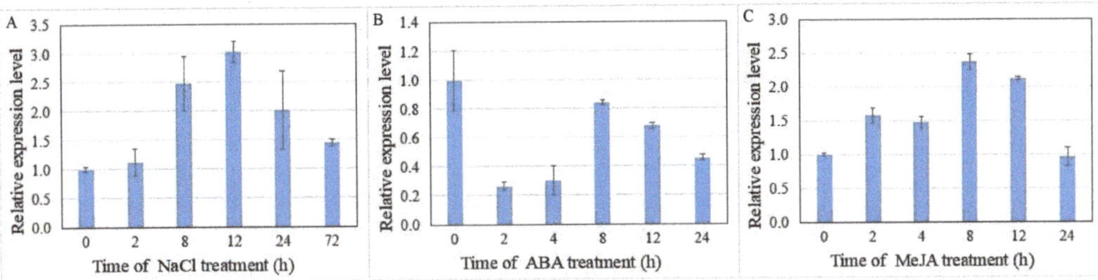

Figure 7. Relative expression analysis of *IrUGT86A1-like* gene at various time periods after abiotic treatments with NaCl (**A**), ABA (**B**), and MeJA (**C**). Results presented by mean value of triplicate ± SE.

3.7. Prokaryotic Expression Analysis of IrUGT86A1-like Protein in I. rubescens

With *E. coli* containing pEASY-Blunt-IrUGT86A1 vector as a template, the target fragment was amplified using IrUGT-YF and IrUGT-YR primers and cloned into the prokaryotic expression vector pEASY®-Blunt E1 Expression Vector. PCR gel electrophoresis was performed using primers IrUGT-YF and T7 Terminator. The results demonstrated that the vector which completed the correct direction of connection and transformed into *E. coli* T1 competent culture could amplify a band of 1500 bp by PCR. The bacterial culture of the vector with incorrect connection direction or failure connection could only amplify a band of less than 100 bp or no band, and only the correct band was sent to sequencing verification. The results showed that the prokaryotic expression vector pEASY-Blunt E1-IrUGT86A1 was successfully constructed (Figure 8A). The successful recombinant plasmid pEASY-Blunt E1-IrUGT86A1 was further transformed into BL21 (DE3) prokaryotic expression competent cells.

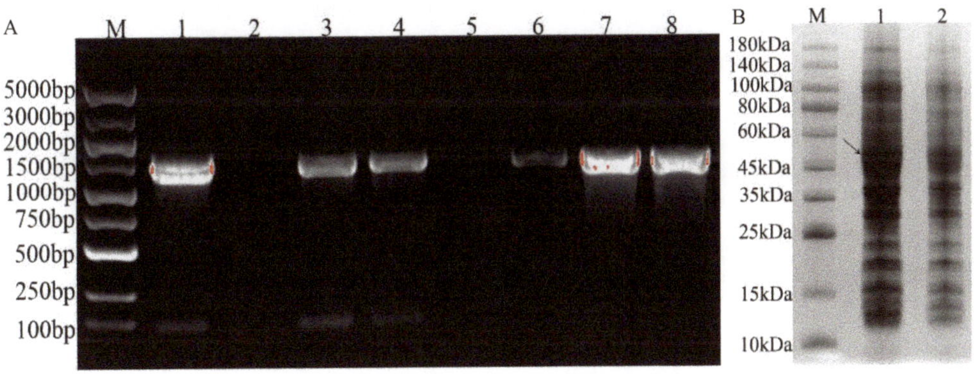

Figure 8. PCR identification and SDS-PAGE analysis of recombinant protein of IrUGT86A1-like. (**A**) PCR identification of recombinant plasmid of IrUGT86A1 (M: DL5000 DNA Maker; 1~8: PCR identification of pEASY-Blunt E1-IrUGT86A colony); (**B**) SDS-PAGE analysis of IrUGT86A recombinant protein (M: Standard molecular weight of protein; 1: pEASY-Blunt E1-IrUGT86A induced by IPTG; 2: pEASY-Blunt E1-IrUGT86A was not induced by IPTG); The arrow indicate the target IrUGT86A1-like protein.

Furthermore, prokaryotic cells containing the target fragment were induced and expressed. The results of protein electrophoresis are shown in Figure 8B. Consistent with predicted results, the target band of *E. coli* strain induced by IPTG was approximately 53 kDa. Compared with the strain without IPTG induction, the same band was not obviously apparent (Figure 8B, Figure S1). Therefore, the IrUGT86A1-like protein was successfully expressed in *E. coli* BL21, which thus lays an important foundation for further study of the IrUGT86A1-like protein's modification characteristics.

4. Discussion

Secondary metabolites play a major role in the plant response to adversity stresses. In plants, for the most part of secondary metabolites accumulate as glycoconjugates. Glycosylation is one of the most common modifications of secondary metabolites, and is carried out by enzymes called glycosyltransferases. In addition, glycosylation on small molecular metabolites modulates various biological events in plants [10,24]. At present, glycosyltransferase genes with various functions have been cloned, and the corresponding enzyme activation and related functions have been verified [9–20]. However, due to the specificity of secondary metabolites in different plants, the functions of glycosyltransferases in different species are species-specific. Therefore, the cloning of glycosyltransferase genes from different species is still a hot topic in the field of secondary metabolic molecules in plants.

In this study, we used transcriptome analysis to screen a *UGT* gene, which had significantly higher expression in the leaves (high oridonin content) than the callus (extremely low oridonin content), and speculated that this gene may be involved in the regulation of oridonin synthesis. Based on the sequence of second-generation transcriptome data, the cDNA of *IrUGT86A1-like* with 1450 bp in length, containing a 1440 bp open reading frame (ORF), which encodes for 479 amino acids was amplified by qPCR. Bioinformatics analysis indicating that the protein encoded by *IrUGT86A1-like* gene is a stable hydrophilic protein with several transmembrane domains. It is speculated that IrUGT86A1-like protein is located in the cytoplasm, which is consistent with previously reported results that glycosylation modification mainly occurs in the endoplasmic reticulum and Golgi apparatus [25]. ExPASy analysis showed that IrUGT86A1-like protein has a molecular weight of 53.03 kD, and the IrUGT86A1-like protein approximately 53 kDa was successfully verified and expressed in *E. coli* BL21, which thus lays an important foundation for further study of the IrUGT86A1-like protein's modification characteristics.

The blastp comparison results from NCBI showed that the cloned *IrUGT86A1-like* gene had high homology (query coverage greater than 95%) with UDP-glycosyltransferase 86A1-like genes in the database. In addition, conserved domains analysis in NCBI demonstrated that the protein had GT1_Gtf-like site-specific and non-specific sites of PLN02448 and UDPGT, which indicted that IrUGT86A1-like belongs to the Glycosyltransferase_GTB-type Superfamily. Phylogenetic analysis revealed that IrUGT86A1-like protein is most closely genetically similar to *Salvia splendens*. Thus, it is believed that *IrUGT86A1-like* gene belong to the *UGT86* gene subfamily.

Previous studies have shown that glycosyltransferases are involved in a wide range of biological processes essential to plant life, particularly in plant secondary metabolic pathways where they often play an important role in the modification of secondary metabolites to synthesize final compound that the *UGT86C11* acts as a novel plant UDP-glycosyltransferase involved in the products [8,26]. For example, during the synthesis of flavonoids, the glycosylation catalyzed by UDP-glycosyltransferase is a key step at the end of the flavonoid biosynthesis pathway [20,24]. Additionally, it has been shown that glycosylation can improve the stability, diversity, and biological activity of flavonoids [26]. In particular, Payal Srivastavaa et. Al. found synthesis of neoandrographolide and andrograpanin, which belong to the labdane diterpenes [27]. In our study, these results of *IrUGT86A1-like* expression and oridonin content in of *I. rubescens* showed that the content of oridonin in different tissues varied with the expression level of the *IrUGT86A1-like* gene in different tissues. However, although no oridonin content was detected in roots, the *IrUGT86A1-like* gene was expressed in roots. This implies that perhaps *IrUGT86A1-like* gene plays roles in other functions in roots. In addition, congruent relationship of *IrUGT86A1-like* expression and oridonin content are consistent with previous laboratory studies in which the leaves (high oridonin content) had significantly higher expression of *IrUGT86A1-like* than the callus (extremely low oridonin content). In summary, these results suggest that the *IrUGT86A1-like* gene may be involved in the regulation of oridonin synthesis. Furthermore, based on homologous genes that have similar functional properties, the

IrUGT86A1-like gene, as one of the members of the *UGT86* subfamily, may also be involved in diterpene biosynthesis. Therefore, based on the results of this study, it is speculated that *IrUGT86A1-like* gene may play a regulatory role in the synthesis of oridonin. However, whether *IrUGT86A1-like* gene plays a modification role in oridonin synthesis or not requires further investigation.

In addition, expression of the *UGT* gene has been reported increase in plants subjected to various stress conditions [28–30]. Previous studies on *Arabidopsis thaliana* have shown that the expression of *UGT79B2* and *UGT79B3*, could be highly induced by various abiotic stressors such as cold, salt, and drought. Additionally, *UGT75C1*, *UGT78D2*, *UGT79B2*, and *UGT79B3* in *A. thaliana* can make plants more resistant or more sensitive to stress through glycosylation of flavonoids [31]. In addition, previous reports on the *UGT86A1* gene in *A. thaliana UGT86A1* knockout lines and overexpression lines found that the glycosyltransferase gene plays an important role in enhancing the adaptation of plants to abiotic stress by altering the expression of stress-related genes [32]. In our study, it also indicated that *IrUGT86A1-like* gene also played a certain regulatory role in stress regulation. At the same time, abiotic stressors also affect the accumulation of secondary metabolites [33]. Therefore, *IrUGT86A1-like* gene may also be involved in the regulation of oridonin synthesis under the influence of different stresses. Whether the change of *UGT* gene expression under different stressors has a specific mechanistic relationship with the shifts in oridonin abundance needs further verification.

5. Conclusions

In this study, the *IrUGT86A1-like* gene of *I. rubescens* has been isolated with RT-PCR and characterized with RNA expression analysis, bioinformatics, and prokaryotic expression. *IrUGT86A1-like* gene encodes for a protein with stable and hydrophilic characteristics, containing a transmembrane domain, and located in the cytoplasm. According to RT-qPCR results, the patterns in expression levels of the *IrUGT86A1-like* gene in different tissues were similar to the distribution of oridonin content in the different tissues of *I. rubescens*, preliminary evidence that *IrUGT86A1-like* gene may be involved in the regulation or modification of oridonin synthesis. Moreover, the expression of *IrUGT86A1-like* gene was found to be altered by abiotic stressors (NaCl, ABA, and MeJA treatments). Expression *IrUGT86A1-like* gene was found to be up-regulated by NaCl and MeJA treatments, and down-regulated by ABA treatment. Finally, prokaryotic expression of IrUGT86A1-like protein successfully expressed the approximately 53 kD IrUGT86A1-like protein, thus contributing an important foundation for further functional research of IrUGT86A1-like protein. Overall, the results of this study provide a theoretical basis for elucidating the potential function of the *IrUGT86A1-like* gene in *I. rubescens*, and enabling future investigation of the function of this gene in the synthesis and modification of secondary metabolites.

Supplementary Materials: The following supporting information can be downloaded at: https://www.mdpi.com/article/10.3390/life12091334/s1, Figure S1: Original Western bolt images of pEASY-Blunt E1-IrUGT86A induced by IPTG (Figure 8B). The marker from top to bottom are 180 kDa, 140 kDa, 100 kDa, 80 kDa, 60 kDa, 45 kDa, 35 kDa, 25 kDa, 15 kDa, 10 kDa.; Table S1: All primers used in this paper.; Table S2: Physicochemical properties and localization analysis of IrUGT86A1-like protein.

Author Contributions: Conceptualization, C.L. and S.C.; methodology, C.L. and J.L. (Jinxu Lan); validation, J.L. (Jinxu Lan), B.Z. and H.Y.; formal analysis, K.G. and J.L. (Jingjing Li); investigation, C.L. and J.L. (Jinxu Lan); resources, H.Y.; data curation, J.L. (Jinxu Lan); writing—original draft preparation, C.L.; writing—review and editing, C.L. and B.Z.; funding acquisition, S.C. All authors have read and agreed to the published version of the manuscript.

Funding: This work was supported by grants from "Henan Province Chinese herbal medicine industry technology system, No.14 [2018]" and "Fourth National Traditional Chinese Medicine Resource Survey Project, No.40 [2019]".

Institutional Review Board Statement: Not applicable.

Informed Consent Statement: Not applicable.

Data Availability Statement: Not applicable.

Conflicts of Interest: The authors declare no conflict of interest.

References

1. National Pharmacopoeia Commission of the People's Republic of China. *Pharmacopoeia of the People's Republic of China*; China Medical Science Press: Beijing, China, 2020.
2. Leung, C.H.; Shiah, H.S.; Liu, S.H.; Jiang, Z.; Gao, W.; Lam, W.; Grill, S.P.; Han, Q.B.; Sun, H.D.; Cheng, Y.C. Exploration of a Chinese herb, *Isodon rubescens*, for cancer treatment with focus on modulation of nuclear factor-kappa B activity. *Cancer Res.* **2004**, *64*, 101–111.
3. Li, D.; Han, T.; Xu, S.; Zhou, T.; Tian, K.; Hu, X.; Cheng, K.; Li, Z.; Hua, H.; Xu, J. Antitumor and Antibacterial Derivatives of Oridonin: A Main Composition of Dong-Ling-Cao. *Molecules* **2016**, *21*, 575. [CrossRef] [PubMed]
4. Meade-Tollin, L.C.; Wijeratne, E.M.; Cooper, D.; Guild, M.; Jon, E.; Fritz, A.; Zhou, G.X.; Whitesell, L.; Liang, J.Y.; Gunatilaka, A.A. Ponicidin and oridonin are responsible for the antiangiogenic activity of *Rabdosia rubescens*, a constituent of the herbal supplement PC SPES. *J. Nat. Prod.* **2004**, *67*, 2–4. [CrossRef] [PubMed]
5. Jin, B.; Cui, G.; Guo, J.; Tang, J.; Duan, L.; Lin, H.; Shen, Y.; Chen, T.; Zhang, H.; Huang, L. Functional Diversification of Kaurene Synthase-Like Genes in *Isodon rubescens*. *Plant Physiol.* **2017**, *174*, 943–955. [CrossRef] [PubMed]
6. Wei, Y.; He, D.; Wei, Y. Study on Soil and Water Conservation Benefits of Herba Rubescentis. *J. Soil Water Conserv.* **2000**, *14*, 24–27. (In Chinese)
7. Strasser, R.; Seifert, G.; Doblin, M.S.; Johnson, K.L.; Ruprecht, C.; Pfrengle, F.; Bacic, A.; Estevez, J.M. Cracking the "Sugar Code": A Snapshot of N-and O-Glycosylation Pathways and Functions in Plants Cells. *Front. Plant Sci.* **2021**, *12*, 640919. [CrossRef]
8. Strasser, R. Plant protein glycosylation. *Glycobiology* **2016**, *26*, 926–939. [CrossRef]
9. Gachon, C.M.; Langlois-Meurinne, M.; Saindrenan, P. Plant secondary metabolism glycosyltransferases: The emerging functional analysis. *Trends Plant Sci.* **2005**, *10*, 542–549. [CrossRef]
10. Tiwari, P.; Sangwan, R.S.; Sangwan, N.S. Plant secondary metabolism linked glycosyltransferases: An update on expanding knowledge and scopes. *Biotechnol. Adv.* **2016**, *34*, 714–739. [CrossRef]
11. Yonekura-Sakakibara, K.; Hanada, K. An evolutionary view of functional diversity in family 1 glycosyltransferases. *Plant J.* **2011**, *66*, 182–193. [CrossRef]
12. Kumar, R.; Sangwan, R.S.; Mishra, S.; Sabir, F.; Sangwan, N.S. In silico motif diversity analysis of the glycon preferentiality of plant secondary metabolic glycosyltransferases. *Plant Omics* **2012**, *3*, 200–210.
13. Rehman, H.M.; Nawaz, M.A.; Shah, Z.H.; Ludwig-Muller, J.; Chung, G.; Ahmad, M.Q.; Yang, S.H.; Lee, S.I. Comparative genomic and transcriptomic analyses of Family-1 UDP glycosyltransferase in three *Brassica* species and *Arabidopsis* indicates stress-responsive regulation. *Sci. Rep.* **2018**, *8*, 1875. [CrossRef] [PubMed]
14. Zhu, T.; Liu, H.; Wang, P.; Ni, R.; Sun, C.; Yuan, J.; Niu, M.; Lou, H.; Cheng, A. Functional characterization of UDP-glycosyltransferases from the liverwort *Plagiochasma appendiculatum* and their potential for biosynthesizing flavonoid 7-O-glucosides. *Plant Sci.* **2020**, *299*, 110577. [CrossRef] [PubMed]
15. Aoi, Y.; Hira, H.; Hayakawa, Y.; Liu, H.; Fukui, K.; Dai, X.; Tanaka, K.; Hayashi, K.; Zhao, Y.; Kasahara, H. UDP-glucosyltransferase *UGT84B1* regulates the levels of indole-3-acetic acid and phenylacetic acid in *Arabidopsis*. *Biochem. Biophys. Res. Commun.* **2020**, *532*, 244–250. [CrossRef] [PubMed]
16. Yonekura-Sakakibara, K.; Fukushima, A.; Nakabayashi, R.; Hanada, K.; Matsuda, F.; Sugawara, S.; Inoue, E.; Kuromori, T.; Ito, T.; Shinozaki, K.; et al. Two glycosyltransferases involved in anthocyanin modification delineated by transcriptome independent component analysis in *Arabidopsis thaliana*. *Plant J.* **2012**, *69*, 154–167. [CrossRef]
17. Xing, L.; Gao, L.; Chen, Q.; Pei, H.; Di, Z.; Xiao, J.; Wang, H.; Ma, L.; Chen, P.; Cao, A.; et al. Over-expressing a UDP-glucosyltransferase gene (*Ta-UGT3*) enhances *Fusarium* Head Blight resistance of wheat. *Plant Growth Regul.* **2018**, *84*, 561–571. [CrossRef]
18. Dong, N.Q.; Sun, Y.; Guo, T.; Shi, C.L.; Zhang, Y.M.; Kan, Y.; Xiang, Y.H.; Zhang, H.; Yang, Y.B.; Li, Y.C.; et al. UDP-glucosyltransferase regulates grain size and abiotic stress tolerance associated with metabolic flux redirection in rice. *Nat. Commun.* **2020**, *11*, 2629. [CrossRef]
19. Behroozi, P.; Baghizadeh, A.; Saei, A.; Kharazmi, S. Quantitative analysis of uridine diphosphate glycosyltransferase *UGT85C2*, *UGT74G1* and *UGT76G1* genes expression in *Stevia rebaudiana* under different irrigations. *Russ. J. Plant Physiol.* **2017**, *64*, 67–72. [CrossRef]
20. Li, J.; Liu, X.; Gao, Y.; Zong, G.; Wang, D.; Liu, M.; Fei, S.; Wei, Y.; Yin, Z.; Chen, J.; et al. Identification of a UDP-Glucosyltransferase favouring substrate- and regio-specific biosynthesis of flavonoid glucosides in *Cyclocarya paliurus*. *Phytochemistry* **2019**, *163*, 75–88. [CrossRef]
21. Yang, Y.; Wang, C.; Wei, M. A green approach for the extraction and characterization of oridonin and ursolic and oleanolic acids from *Rabdosia rubescens* and its kinetic behavior. *Food Chem.* **2020**, *319*, 126582. [CrossRef]
22. Kumar, S.; Stecher, G.; Tamura, K. MEGA7: Molecular Evolutionary Genetics Analysis Version 7.0 for Bigger Datasets. *Mol. Biol. Evol.* **2016**, *33*, 1870–1874. [CrossRef] [PubMed]

23. Livak, K.J.; Schmittgen, T.D. Analysis of relative gene expression data using real-time quantitative PCR and the 2(-Delta Delta C(T)) Method. *Methods* **2001**, *25*, 402–408. [CrossRef] [PubMed]
24. Campos, L.; López-Gresa, M.P.; Fuertes, D.; Bellés, J.M.; Rodrigo, I.; Lisón, P. Tomato glycosyltransferase Twi1 plays a role in flavonoid glycosylation and defence against virus. *BMC Plant Biol.* **2019**, *19*, 450. [CrossRef]
25. Schoberer, J.; Strasser, R. Plant glyco-biotechnology. *Semin. Cell Dev. Biol.* **2018**, *80*, 133–141. [CrossRef] [PubMed]
26. Le Roy, J.; Huss, B.; Creach, A.; Hawkins, S.; Neutelings, G. Glycosylation Is a Major Regulator of Phenylpropanoid Availability and Biological Activity in Plants. *Front. Plant Sci.* **2016**, *7*, 735. [CrossRef]
27. Srivastava, P.; Garg, A.; Misra, R.C.; Chanotiya, C.S.; Ghosh, S. *UGT86C11* is a novel plant UDP-glycosyltransferase involved in labdane diterpene biosynthesis. *J. Biol. Chem.* **2021**, *297*, 101045. [CrossRef]
28. Pan, L.; Yan, J.L.; Bo, W.; Hui, M.Y.; Qin, L.; Bing, K.H. The *Arabidopsis UGT87A2*, a stress-inducible family 1 glycosyltransferase, is involved in the plant adaptation to abiotic stresses. *Physiol. Plant.* **2017**, *159*, 416–432.
29. Wu, C.; Dai, J.; Chen, Z.; Tie, W.; Yan, Y.; Yang, H.; Zeng, J.; Hu, W. Comprehensive analysis and expression profiles of cassava UDP-glycosyltransferases (UGT) family reveal their involvement in development and stress responses in cassava. *Genomics* **2021**, *113*, 3415–3429. [CrossRef]
30. Li, Y.; Liu, F.; Li, P.; Wang, T.; Zheng, C.; Hou, B. An arabidopsis Cytokinin-Modifying glycosyltransferase *UGT76C2* improves drought and salt tolerance in rice. *Front. Plant Sci.* **2020**, *11*, 560696. [CrossRef]
31. Li, P.; Li, Y.J.; Zhang, F.J.; Zhang, G.Z.; Jiang, X.Y.; Yu, H.M.; Hou, B.K. The *Arabidopsis* UDP-glycosyltransferases *UGT79B2* and *UGT79B3*, contribute to cold, salt and drought stress tolerance via modulating anthocyanin accumulation. *Plant J.* **2017**, *89*, 85–103. [CrossRef]
32. Zhang, F.J. Functional Analysis of *Arabidopsis* Glycosyltransferase Gene *UGT86A1* in Plant Adaptation to Abiotic Stresses. Master's Thesis, Shandong University, Jinan, China, 2018.
33. Zandalinas, S.I.; Balfagón, D.; Gómez-Cadenas, A.; Mittler, R. Plant responses to climate change: Metabolic changes under combined abiotic stresses. *J. Exp. Bot.* **2022**, *73*, 3339–3354. [CrossRef] [PubMed]

life

MDPI

Article

Effects of Different Growth Regulators on the Rooting of *Catalpa bignonioides* Softwood Cuttings

Jin'e Quan †, Ruoyi Ni †, Yange Wang, Jiajia Sun, Mingyue Ma and Huitao Bi *

College of Forest, Henan Agricultural University, Zhengzhou 450002, China
* Correspondence: bihuitao@126.com; Tel.: +86-177-4462-4393
† These authors contributed equally to this work.

Abstract: (1) Background: To further improve the rapid reproduction and large-scale application of *Catalpa bignonioides*. (2) Methods: With young softwood cuttings from a 3-year-old *C. bignonioides* mother plant used as materials, the effects of indole-3-acetic acid(IAA), indolebutyric acid(IBA) and rhizogenic powder-1(ABT-1) growth regulators at different concentrations on cutting indexes and the dynamic changes in endogenous hormone contents during the rooting of the *C. bignonioides* cuttings were studied. (3) Results: The rooting of *C. bignonioides* cuttings could be divided into five stages. There were three types of rooting of adventitious roots. IBA treatment resulted in a high rooting rate and beneficial root morphology. The morphological indexes of the cutting roots after treatment with 1000 mg·L^{-1} IBA had the best overall quality, which was significantly higher than that of the roots in the control (CK) group ($p < 0.05$). Although the average longest root length (20.51 cm) under ABT-1 was the longest, its overall average rooting rate was slightly lower than that under IBA. The rooting effect under IAA was generally lower than that under IBA and ABT-1. The endogenous hormone content of the cuttings was found to be closely related to rooting; the IAA and zeatin nucleoside (ZR) content was high, and the ratios of IAA/ABA and IAA/ZR were high. The contents of gibberellin$_3$ (GA_3) and abscisic acid (ABA) were low, which had a promoting effect on the rooting of the cuttings. (5) Conclusions: All three kinds of auxin can promote rooting and, of the three treatment groups, the rooting effect of cuttings in the IBA treatment group was the strongest, with 1000 mg·L^{-1} being the optimum concentration.

Keywords: *Catalpa bignonioides*; softwood cutting; growth hormone; rooting index

Citation: Quan, J.; Ni, R.; Wang, Y.; Sun, J.; Ma, M.; Bi, H. Effects of Different Growth Regulators on the Rooting of *Catalpa bignonioides* Softwood Cuttings. *Life* **2022**, *12*, 1231. https://doi.org/10.3390/life12081231

Academic Editors: Jie Luo and Sen Meng

Received: 3 July 2022
Accepted: 4 August 2022
Published: 15 August 2022

1. Introduction

Catalpa bignonioides, also known as cigar tree, American catalpa and Indian bean tree, is a deciduous tree species of the family Bignoniaceae and is native to North America. *C. bignonioides* is grown mainly in parts of Canada and the central and southern United States. In recent years, *C. bignonioides* has also been introduced in the northern region of China to central Xinjiang, Heilongjiang, Jilin, Liaoning, and south to Yunnan and other provinces [1]. Owing to its strong vigor, large leaves, and floral fragrance, as well as its ability to produce dense shade, *C. bignonioides* is mainly used as garden ornamental or street tree. Its flowers are white and bell-shaped, with yellow or lavender spots. The pods are similar to long beans and hang throughout the tree canopy, which is very attractive. There are different leaf colors, including green, yellow, and purple, as well as multicolored leaves. Attractive features of these trees are visible in three seasons, namely, its leaves in the spring, its flowers in the summer, and its fruits in autumn. The leaves are large and rough and have the ability to strongly absorb particles and heavy metals, such as lead and cadmium. Owing to its broad canopy, *C. bignonioides* is often used as a wind barrier [2]. Wei Zuoping et al. [3] reported that *C. bignonioides* is an important primary landscaping tree species with optimal landscape effects. Bi Huitao et al. [4] treated softwood cuttings of *C. bignonioides* with different concentrations of IBA, and its rooting rate increased by 84% compared with that of the cuttings in the CK group. In recent years, although research on *C. bignonioides* has mainly focused on trees of the same family and genus, relatively few studies have focused on seedling propagation [4]. Chen Suchuan et al. [5] and Liang Youwang et al. [6] conducted preliminary research on the sowing of *Catalpa* species and the rooting of softwood cuttings, and Zhao Xiyang et al. [7] and Peng Chan et al. [8] explored the seed germination characteristics of *Catalpa eucalyptus* and the effects of exogenous hormones on the growth of *Eucalyptus* seedlings. Cui Lingjun et al. [9] carried out subjected *Catalpa bungei* seeds to different hormone treatments and concluded that IBA promotes seed germination. However, the speed of propagation via seed is slow due to the dormancy characteristics of these species [3]. In addition, limited domestic seed resources prevent rapid and large-scale promotion and application. The cutting propagation technique has advantages that include the stable maternal inheritance of excellent traits, a short growth cycle, ease of operation, and rapid prototyping; thus, propagation by cuttings has been considered to be the most cost-effective reproduction method in recent years and can hasten and promote the propagation of large numbers of seedlings [10]. Therefore, the cutting propagation technique is beneficial to accelerate the promotion and application of *C. bignonioides* reproduction.

At present, there has been no research on *C. bignonioides* cutting propagation techniques in China and other countries. The low rooting rate of cuttings has been a major bottleneck in forestry development in China. To overcome this limitation, many researchers have carried out related research on the effects of growth regulators on various species, such as *Chionanthus virginicus, Lonicera korolkowi, Quercus mongolica, Mytilaria laosensis*, and other tree species. According to the research results, the rooting rate of cuttings treated with exogenous growth regulators was found to be significantly higher than that of untreated cuttings [11–14]. The selection and application of auxin is also a key factor for improving the rooting rate of cuttings. In addition, treatment with growth regulators whose concentrations are too high or too low is not conducive to the rooting of cuttings [15] because low or high concentrations of exogenous growth regulators can disrupt the balance of endogenous hormone contents within the cutting itself, resulting in an inability to promote adventitious root production [16]. Wang Qing et al. used *Chukrasia tabularis* shoots cultured in MS-containing medium as the research object and found that adding IBA and ABT-1 at a concentration of 500 mg·L^{-1} induced rooting after 10 days. Furthermore, IBA processing outperforms ABT-1 processing, with a rooting rate over 85% [17]. By using three kinds of growth regulators (naphthaleneacetic acid (NAA), IBA, and IAA) at different concentrations and treatment times, Zhai Yafang et al. conducted a cutting propagation test on *Lonicera tartarii* cuttings. According to the test results, the rooting rate under IBA 500 mg·L^{-1} was the best at 86% [18]. Wang Xiaoling et al. soaked tetraploid *Robinia pseudoacacia* softwood in different concentration gradients (500 mg·L^{-1}, 1000 mg·L^{-1}, 1500 mg·L^{-1}) of different hormones (IBA, NAA, and IAA) for 6 h. The best treatment group was IBA at a concentration of 1000 mg·L^{-1} with a rooting rate of 80.4% [19]. Zhang Enliang treated *Catalpa* softwood with IBA at a concentration of 2000 mg·L^{-1}, with a rooting rate of 85.6% [20]. Ma Lingling et al. evaluated the cutting rooting ability of five species of *Catalpa* softwood. After soaking in IBA with a concentration of 3000 mg·L^{-1} for 1 min, the rooting rate of *Catalpa* was as high as 94.5% [21]. Wang Gaiping used NAA, IAA, and ABT-1 at a concentration of 1000 mg·L^{-1} to analyze the rooting development and root characteristics of the cuttings of *Catalpa* softwood. Among them, the ABT-1 treatment group with a concentration of 1000 mg·L^{-1} was the best with a rooting rate of 92% [22]. Zhang Mei et al. used different concentrations of ABT-1, NAA, and IAA to treat 1-year-old cuttings of *Brassica chinensis*. Among them, the rooting rate of ABT-1 with a concentration of 500 mg·L^{-1} was the highest, followed by the IAA treatment group [23]. In order to determine a suitable cutting plan for cold-resistant plum root stocks, Wang Xuesong et al. soaked the rooting powder in NAA, IAA, IBA, and ABT-1 at concentrations of 1, 3, and 5 g·L^{-1} for 40 min. The best treatment group was ABT-1 at a concentration of 3 g·L^{-1} [24]. Liang Xiaochun et al. used different concentrations of the exogenous hormones IAA and IBA to treat Huangguohougui softwood and found that the IAA treatment group with a concentration of 250 mg·L^{-1}

was the best, with a rooting rate of 66.7% [25]. Li Huimin et al. studied the effect of different concentrations of plant growth regulators (IBA, IAA, and NAA) on the rooting survival rate of cuttings of original varieties of perfume rose. Among them, the survival rate of cuttings treated with IAA solution with a concentration of 500 mg·L^{-1} was the highest [26]. Wu Kaiyun et al. used various concentrations of plant growth regulators (IBA, ABT-1, NAA, shuang ji er-GGR6, and IAA) to soak hardwood cuttings from the persimmon rootstock Yalin 6 and then inserted them into a sterile matrix. The highest rooting rate was found for the IAA treatment group with a concentration of 500 mg·L^{-1}, followed by the ABT-1 treatment group [27]. The best growth regulators widely used to rapidly increase the rooting rate of cuttings are IBA, ABT-1, and IAA, which are suitable at concentrations ranging from 500 mg·L^{-1} to 2000 mg·L^{-1}.

According to a large number of studies in recent years, the changes in the contents of endogenous hormones during the rooting process of cuttings are more important for regulating the occurrence of adventitious roots than their contents in a specific developmental period. In addition, treatment with growth regulators whose concentrations are too high or too low is not conducive to the rooting of cuttings [28,29]. In a study by Zhang Xiaoping et al. [30] on *Liriodendron chinense* × L. *tulipifera* and a study by Li Chaochan et al. [31] on *Rhododendron stamineum*, IAA was found to promote the formation of adventitious roots and ABA had a certain inhibitory effect on root formation. Exogenous hormone treatment can promote the synthesis of endogenous IAA and inhibit the synthesis and transport of endogenous ABA. In research on *Rhododendron scabrifolium* [32] the increase in the GA_3 content had a positive correlation with the induction of cutting calli and the occurrence of adventitious roots. According to an experiment on *Swida wilsoniana* by Li Yongxin et al. [33] a high concentration of ZR is beneficial to the growth and development of root primordia, while a low concentration of ZR is beneficial to the formation of root primordia. In research by Dong Shengjun et al. [34] on *Armeniaca sibirica* and by Shi Fenghou et al. [35] on *Tilia miqueliana*, plant growth regulators could affect the changes in endogenous hormones in cuttings and promote cutting rooting.

With the use of young softwood cuttings of 3-year-old *C. bignonioides* mother plants as materials, this experiment involved applying three different growth regulators, namely, IAA, ABT-1, and IBA, at concentrations of 500 mg·L^{-1}, 1000 mg·L^{-1}, and 1500 mg·L^{-1} on the materials to study their impact on the rooting index of the cuttings. The cuttings displaying optimal rooting across the treatment groups were screened, and the dynamic changes in the contents of four main endogenous hormones IAA, ABA, GA_3, and ZR and their ratios during the rooting process were measured. This study identified the best growth regulator to deal with the rooting rate of cuttings and the survival rate of *C. bignonioides* softwood for the first time and provided a theoretical basis for future studies. From the perspective of physiology and biochemistry, the rooting mechanism was explained and can be used for the establishment of a rapid propagation technology system of *C. bignonioides* seedling cuttings and the rapid large-scale production of seedlings with excellent genetic traits. Moreover, technical support and theoretical guidance on the industrialization development of *C. bignonioides* were provided.

2. Materials and Methods

2.1. Test Materials, Cutting Treatments, and Sample Collection

In this research, the materials were softwood cuttings of *C. bignonioides* collected in mid-July from a 3-year-old mother plant at an experimental station in Ruzhou city, Henan Province. The mother plants were all similar in terms of their height. The cuttings were collected from vigorously growing, young branches and were 0.6–0.8 cm in diameter, after which they were all pruned to a length of 12–15 cm. The upper end of each cutting was cut flat, and the base incision was beveled at 45°.

Three growth regulators, ABT-1(Beijing EbTY Biotechnology Co., Ltd., Beijing, China), IAA, and IBA (Beijing Solaibao Technology Co., Ltd., Beijing, China) were applied separately. IAA, IBA, and ABT-1 at 0.5 g, 1 g, and 1.5 g, respectively, were dissolved with 100%

alcohol and then diluted with 1 L of water to a concentration of 500, 1000, and 1500 mg·L^{-1} to soaking 2–3 cm of the cutting's base for 1 h. There were 9 treatments, each of which was replicated 3 times. A total of 80 cuttings constituted a bundle (one replication). Cuttings soaked in water for 1 h were used as CKs, as shown in Table 1.

Table 1. Orthogonal design of different growth regulator treatments. IAA: Indole-3-acetic acid; IBA: Indolebutyric acid; ABT-1: Rhizogenic powder-1; T1: IAA 500 mg·L^{-1} 1 h; T2: IAA 1000 mg·L^{-1} 1 h; T3: IAA 1500 mg·L^{-1} 1 h; T4: IBA 500 mg·L^{-1} 1 h; T5: IBA 1000 mg·L^{-1} 1 h; T6: IBA 1500 mg·L^{-1} 1 h; T7: ABT-1 500 mg·L^{-1} 1 h; T8: ABT-1 1000 mg·L^{-1} 1 h; T9: ABT-1 1500 mg·L^{-1} 1 h; CK: Control.

Treatment Number	A Growth Regulators	B Concentration (mg·L^{-1})	C Treatment Time (h)
T1	IAA	500	1
T2	IAA	1000	1
T3	IAA	1500	1
T4	IBA	500	1
T5	IBA	1000	1
T6	IBA	1500	1
T7	ABT-1	500	1
T8	ABT-1	1000	1
T9	ABT-1	1500	1
CK	Water		

For each bundle of 80 cuttings, the base was first soaked in 50% carbendazol wettable powder (Jiangsu Lanfeng Bio-Chemical Co., Ltd., Nanjing, China) diluted 800 times with water for 1 min. To accurately observe the morphological changes in the base of the cuttings, three cuttings were randomly selected from each treatment every other day for observations, and the rooting status of the cuttings was recorded. Sampling was performed on days 0, 7, 15, 30, and 50 after treatment. Three cuttings were randomly selected for each treatment. First, the cuttings were rinsed with water and dried. Then, the cortex within 2 cm of the cutting was immediately removed. After being cut into pieces, they were wrapped in aluminum foil and put into liquid nitrogen. Finally, the cuttings were stored in a −80 °C ultra-low-temperature freezer in the laboratory for the subsequent determination of endogenous hormone contents.

2.2. Cuttings and Post-Management Processes

In this study, a softwood cutting system used for breeding was adopted. A seedbed with a length of 5 m, a width of 4 m, and a depth of 0.5 m surrounded by bricks was used. The matrix in the seedbed consisted of pure fine river sand:vermiculite:perlite = 3:1:1. An automatic spraying system was installed 1.5 m above the seedbed, and a shading net was placed 10 cm above the automatic spraying system. One week before the cuttings were transplanted, 50% carbendazol wettable powder diluted 800 times with water was evenly sprayed around the seedbed and the substrate. At the same time, the substrate was tumbled. Before the cuttings were transplanted, the seedbed substrate was watered to loosen the matrix particles. The cuttings were transplanted such that the density was 40 m^{-2} and their depth was 7–8 cm. After the cuttings were transplanted, water was sprayed every 30 min from 8:00 to 18:00 every day. To prevent excessive water spraying or high temperatures and water shortages in the summer, a moisture timer controller produced by staff at the Beijing Academy of Forestry was used to monitor the water spraying intervals and times (Chinese Academy of Forestry Sciences, Beijing, China). On sunny days from 10:00 to 16:00, a shade net was placed over the cuttings. The moisture of the cutting substrate was moderate. Moreover, the relative humidity at 1.5 m above the cuttings was maintained at approximately 70%.

2.3. Determination of Rooting Indicators

Beginning on day 5, three cuttings were randomly taken each time to observe the changes in cutting morphological indicators. On day 50, the rooting of all the cuttings in the seedbed was investigated. First, the counting method was used to quantify the callus production rate and rooting rate in each cutting group. Then, an Epson root analyzer (Epson Perfection 4990 Photo scanner, Epson, Nagano, Japan) manufactured in Nagano, Japan, was used to determine the root number, maximum root diameter, and maximum root length of each cutting group.

$$\text{Rooting rate (\%)} = (\text{number of rooted cuttings/total number of cuttings tested}) \times 100\% \quad (1)$$

$$\text{Callus production rate (\%)} = (\text{number of cuttings that produced calli/total number of cuttings}) \times 100\% \quad (2)$$

$$\text{Root number} = \text{sum of root number/total number of roots} \quad (3)$$

$$\text{Average maximum root diameter (mm)} = \text{sum of maximum root diameter/total roots} \quad (4)$$

$$\text{Average maximum root length (cm)} = \text{sum of maximum root length/total number of roots} \quad (5)$$

2.4. Determination of Physiological and Biochemical Indicators

For the determination of physiological indicators of the phloem of *C. bignonioides* cuttings, the method proposed by He Chongdan et al. [29] was used to extract endogenous hormones. Enzyme-linked immunosorbent assays (ELISAs) were used to determine the contents of rooting-related endogenous hormones, including IAA, GA_3, ABA, and ZR, in units of nanograms per gram of fresh weight (FW). Each sample was replicated 3 times.

2.5. Data Processing

SPSS 24.0 software (IBM Corp., Armonk, NY, USA)was used for analysis of variance and correlation analysis. Origin 8 software(Originlab, Northampton, MA, USA) was used for mapping. Duncan's method was used for multiple comparisons, and the mean ± SE was used to represent the test results.

3. Results and Analysis

3.1. Rooting of C. bignonioides Softwood Cuttings

The morphological changes in the *C. bignonioides* cuttings are as follows. Beginning on day 10, some cutting bases produced a small amount of milky calli (Figure 1A). On day 15, the calli gradually increased (Figure 1B) and on day 20, the calli continued to gradually increase (Figure 1C). On day 30, some adventitious roots continued to extend from inside the calli (Figure 1D). A few adventitious roots had been generated from the bark of the cuttings (Figure 1E). Some adventitious roots arose from the interior and cortex of the calli of the same cutting (Figure 1F) and elongated. In addition, new roots had formed on these roots. Finally, a complex interlaced root system formed (Figure 1G). According to the phenotypic changes in the rooting of the *C. bignonioides* cuttings (Figure 1), the process could be divided into five stages, namely, the initiation stage, callus stage, root primordium induction stage, adventitious root generation period, and elongation stage. According to previous research, the different rooting types can be divided into bark rooting types, callus rooting types, and mixed rooting types [16,36]. Figure 1 shows that the rooting process of the *C. bignonioides* cuttings involves these three rooting types.

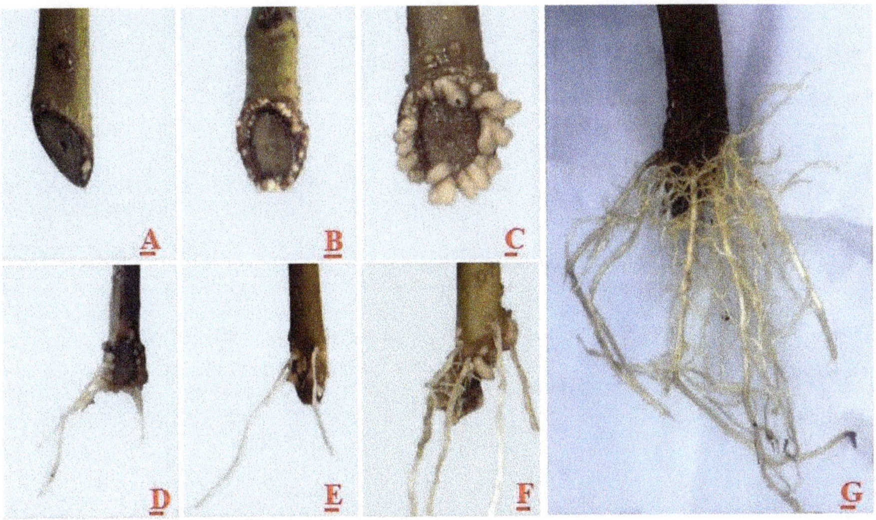

Figure 1. Development process of adventitious roots of *C. bignonioides* cuttings. (**A**) A small amount of callus was generated; (**B**) A large number of calli were generated; (**C**) Callus increase; (**D**) Adventitious roots protruding from the calli; (**E**) Adventitious roots protruding from the bark; (**F**) Calli and bark of the same cutting producing adventitious roots; (**G**) Calli and the bark of the same cutting producing adventitious roots that have elongated.

3.2. Comparison of Rooting Rate and Callus Production Rate in Response to Different Growth Regulators

3.2.1. Influence of Growth Regulator on the Rooting Rate and Callus Production Rate of *C. bignonioides* Softwood Cuttings

Table 2 shows that the rooting rate and callus production rate of *C. bignonioides* cuttings under the three growth regulator treatments were higher than those under the CK treatment, with a consistent trend. The impact of the growth regulators on these two indicators was in the order of IBA > ABT-1 > IAA. Among the cuttings under the three growth regulators, the rooting rate under IBA was 62.44%, which was 7.81 times that under the CK and was significantly greater than that under the other growth regulators. Moreover, the rooting rate in response to ABT-1 was significantly greater than that under IAA. The callus production rate under IBA was 18.44%, which was 9.22 times that under the CK; this production was not significantly different from that under ABT-1 but was significantly greater than that under the other growth regulator treatments. The difference between the IAA and CK treatments was not significant. Overall, the rooting rate under the three growth regulators treatments was greater than the callus production rate, and the IBA treatment elicited the best response.

Table 2. Effects of growth regulators on the rooting rate and callus production rate of *C. bignonioides* softwood cuttings. IAA: Indole-3-acetic acid; IBA: Indolebutyric acid; ABT-1: Rhizogenic powder-1,CK: Control. The data in the table are the mean of 3 replications. The different lowercase letters (a, b, c, and d) after the data in the same column indicate significant differences at the $p < 0.05$ level, and the different capital letters (A, B, C, and D) indicate significant differences at the $p < 0.01$ level.

Growth Regulator	Rooting Rate/%	Callus Production Rate/%
IAA	32.89 ± 1.68 cC	8.22 ± 3.15 bAB
IBA	62.44 ± 2.14 aA	18.44 ± 3.67 aA
ABT-1	49.33 ± 1.76 bB	17.56 ± 5.35 aA
CK	8.00 ± 2.00 dD	2.00 ± 2.00 bB

3.2.2. Effects of Growth Regulator Concentration on the Rooting Rate and Callus Production Rate of *C. bignonioides* Softwood Cuttings

The rooting rate of cuttings treated with different growth regulators is shown in Figure 2A. Overall, the rooting rate of cuttings in the IBA treatment group was the best (43–84%) and was significantly higher than that of the cuttings in the IAA and ABT-1 treatment groups. With an increase in IBA concentration, the rooting rate tended to first increase and then decrease. Among the different concentrations tested, the maximum rate was 84.00% in response to 1000 mg·L^{-1}, which was 90.48% higher than that under the CK (the rooting rate of which was 8.00%). The second greatest rooting rate was detected in the ABT-1 treatment group. With the increase in ABT-1 concentration, the rooting rate also increased first and then decreased. The rooting rate of the cuttings was the best after treatment with 500 mg·L^{-1} ABT-1 compared with other concentrations, which was 86.52% higher than that of the CK. In general, the rooting rate of the IAA treatment group was the lowest (28–38%); IAA had no significant effect on the rooting rate.

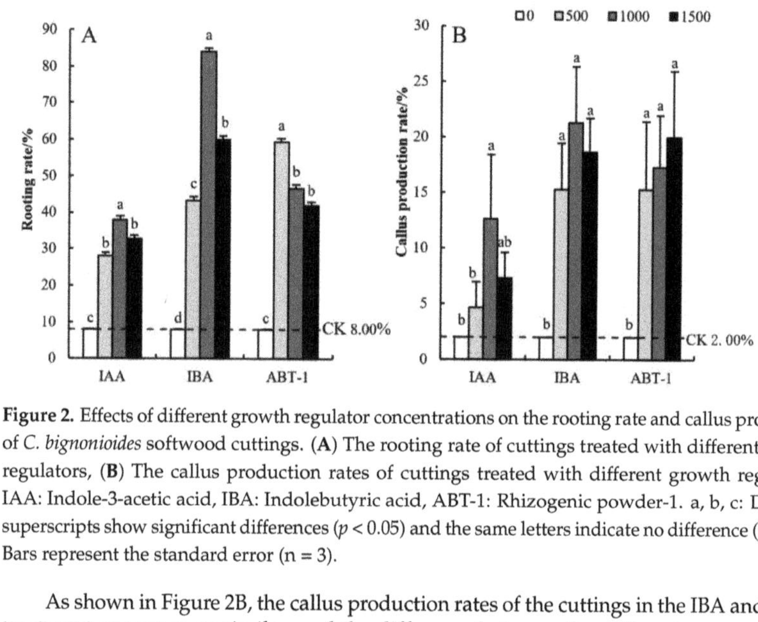

Figure 2. Effects of different growth regulator concentrations on the rooting rate and callus production of *C. bignonioides* softwood cuttings. (**A**) The rooting rate of cuttings treated with different growth regulators, (**B**) The callus production rates of cuttings treated with different growth regulators, IAA: Indole-3-acetic acid, IBA: Indolebutyric acid, ABT-1: Rhizogenic powder-1. a, b, c: Different superscripts show significant differences ($p < 0.05$) and the same letters indicate no difference ($p > 0.05$). Bars represent the standard error (n = 3).

As shown in Figure 2B, the callus production rates of the cuttings in the IBA and ABT-1 treatment groups were similar, and the difference between the different concentrations of hormones was not significant (15–22%); however, the production rates in response to both were significantly higher than that of the CK. Among the treatment groups, the callus production rate of the IBA treatment group first increased and then decreased with increasing concentration, which is consistent with the rooting rate trend shown in Figure 2A. The callus production rate of the cuttings in the ABT-1 treatment group showed a gradually increasing trend with increasing concentration, which is different from the rooting rate trend shown in Figure 2A. The cuttings in the IAA treatment group presented the lowest callus production rate (4–13%), which was lower overall than that in the IBA and ABT-1 treatment groups. In addition, the overall trend and significance between the different hormone concentrations were essentially consistent with those of the rooting rate, as shown in Figure 2A.

3.3. Comparison of Root Morphology in Response to Different Growth Regulator Treatments

3.3.1. Effects of Growth Regulators on Root Morphology of C. bignonioides Softwood Cuttings

The root morphological changes of the *C. bignonioides* softwood cuttings are shown in Table 3. The three growth regulators significantly affected the number of roots produced, average maximum root length, and average maximum root diameter of the cuttings. In the IBA treatment, the number of roots produced reached 16.56, which was 2.37 times that in the CK. This was significantly greater than the numbers in the ABT-1 treatment (10.11 roots) and IAA treatment (8.33 roots) ($p < 0.05$). Second, ABT-1, IBA, and IAA all significantly promoted the average maximum root length of the *C. bignonioides* softwood cuttings. Among these treatments, the promoting effect of ABT-1 was the strongest (20.51 cm on average), which was significantly higher than that of CK (10.40 cm). Finally, the average maximum root diameter in response to the three growth regulators ranged from 1.27–1.61 mm, which equates to the roots being 41.67~58.82% thicker than those of the CK (0.7 mm on average). Therefore, overall, the hormone treatment of *C. bignonioides* softwood cuttings has a significant promoting effect on maximum root diameter. Among the treatments, the effect of IBA was the strongest and was significantly better than that of ABT-1.

Table 3. Effects of growth regulators on the root morphological indicators of *C. bignonioides* softwood cuttings. IAA: Indole-3-acetic acid; IBA: Indolebutyric acid; ABT-1: Rhizogenic powder-1; CK: Control. The data in the table are the mean of three replications. The different lowercase letters (a, b) after the data in the same column indicate significant differences ($p < 0.05$), and the different capital letters (A, B) indicate significant differences ($p < 0.01$).

Growth Regulator	Number of Roots/n	Average Maximum Root Length/cm	Average Maximum Root Diameter/mm
IAA	8.33 ± 2.91 bA	16.34 ± 7.38 abA	1.47 ± 0.28 aA
IBA	16.56 ± 3.42 aA	18.31 ± 0.88 aA	1.61 ± 0.30 aA
ABT-1	10.11 ± 3.79 bA	20.51 ± 0.62 aA	1.27 ± 0.19 aAB
CK	7.00 ± 3.00 bA	10.40 ± 1.13 bA	0.70 ± 0.10 bB

3.3.2. Effects of Growth Regulator Concentration on the Root Morphology of C. bignonioides Softwood Cuttings

As shown in Figure 3A, the number of roots produced by the cuttings in the IBA treatment group was the highest. In addition, the number of roots in response to IBA at concentrations of 1000 mg·L^{-1} and 1500 mg·L^{-1} were significantly higher than those in response to ABT-1 and IAA. The number of roots first increased and then decreased slightly with increasing IBA concentration. The maximum value was 22.00 under the 1000 mg·L^{-1} IBA treatment, which was approximately 2.36 times that under the CK treatment. Therefore, IBA concentrations that are too high or too low are not conducive to the development of adventitious roots. The effects of the ABT-1 and IAA treatments on the number of roots produced were relatively weak (5~14 roots); there was no significant difference in number of roots produced by the cuttings between the concentrations of these hormones or between the CK. Therefore, the promoting effect of ABT-1 and IAA on the number of roots was not obvious. When the IAA and ABT-1 concentrations reached 1500 mg·L^{-1}, root production (7.33 and 5.00) became inhibited, and the number produced was significantly lower than that in the CK treatment (9.33).

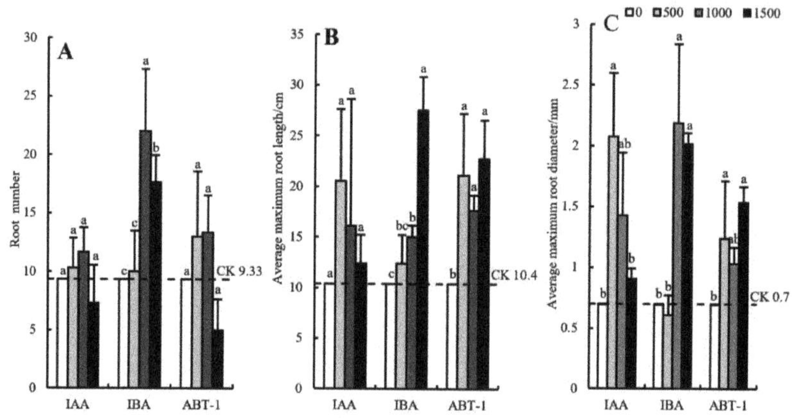

Figure 3. Effects of growth regulator concentrations on the root morphological indexes of *C. bignonioides* softwood cuttings. (**A**) The number of roots produced by the cuttings treated with different growth regulators, (**B**) The average maximum root length by the cuttings treated with different growth regulators, (**C**)The average maximum root diameterby the cuttings treated with different growth regulators, IAA: Indole-3-acetic acid; IBA: Indolebutyric acid; ABT-1: Rhizogenic powder-1. a, b, c: Different superscripts show significant differences ($p < 0.05$) and the same letters indicate no difference ($p > 0.05$). Bars represent the standard error (n = 3).

As shown in Figure 3B, all three growth regulators had a significant promoting effect on the average maximum root length. The maximum root length of the cuttings in the IBA treatment group with a concentration of 1500 mg·L^{-1} was the largest (27.53 cm) among all the treatment groups; this value was significantly higher than that in the other treatment groups (across concentrations) and was 62.22% higher than that in the CK group. Second, the average maximum root length under the ABT-1 treatment fluctuated in the range of 17–23 cm; there was no significant difference between the different concentrations, but the length in response to the different concentrations were significantly higher than that in the CK treatment. The average maximum root length under the IAA treatment first increased and then decreased with increasing IAA concentration. However, there was no significant difference in average maximum root length between the concentrations or the CK.

As shown in Figure 3C, IBA at a concentration of 1000 mg·L^{-1} resulted in the greatest average maximum root diameter (2.18 mm), which was 67.89% thicker than that in the CK group (0.7 mm). The average maximum root diameter of the cuttings in the IBA treatment group with a concentration of 1500 mg·L^{-1} was slightly lower than that in the 1000 mg·L^{-1} IBA treatment group but was significantly higher than that in the 500 mg·L^{-1} and CK treatment groups. The average maximum root diameter under IAA treatment first increased and then decreased with increasing IAA concentration. The average maximum root diameter under the 500 mg·L^{-1} IAA treatment was the greatest (2.07 mm) and was 66.18% thicker than that under the CK. The average maximum root diameter under the ABT-1 treatment fluctuated in the range of 1.03–1.54 mm. Among the various ABT-1 concentrations tested, the average maximum root diameter under the treatment of 1500 mg·L^{-1} ABT-1 was the greatest, the value of which was significantly different from that under the CK. The average maximum root diameter in response to the other concentrations of ABT-1 was not significantly different from that of the CK group.

3.4. Changes in the Contents of Hormones in the Phloem of C. bignonioides Cuttings under Different Treatments

3.4.1. Changes in the Content of IAA in the Phloem of *C. bignonioides* Cuttings

Figure 4A shows that the content of IAA in the phloem of *C. bignonioides* cuttings treated with growth regulators during the rooting process first increased and then decreased with time. Specifically, the IAA content increased on days 0–15 and decreased on days 15–50. Among them, treatment 5 corresponded to the largest increase. On day 15, the IAA content peaked at 122.10 ng·g FW, which was 5.26 times the initial content of the cuttings (23.22 ng·g FW). Moreover, the content was higher than that under the other treatments. The CK treatment yielded the smallest increase, and the value remained the lowest throughout rooting.

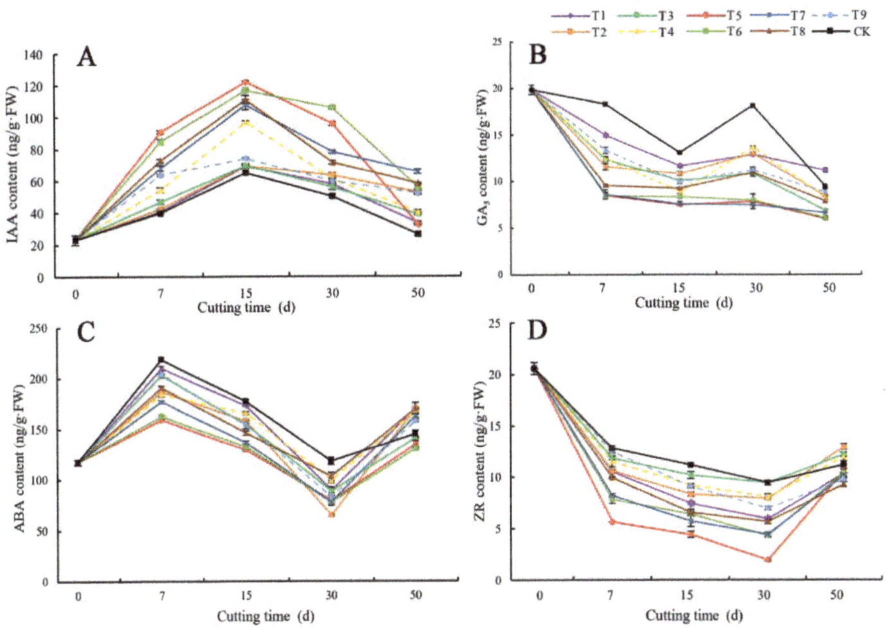

Figure 4. Changes in the content of IAA, GA_3, ABA, and ZR in the phloem of *C. bignonioides* cuttings during rooting. (**A**) Changes in the content of IAA in the phloem of C. bignonioides cuttings during rooting, (**B**) Changes in the content of GA_3 in the phloem of C. bignonioides cuttings during rooting, (**C**) Changes in the content ofABA in the phloem of C. bignonioides cuttings during rooting, (**D**) Changes in the content of ZR in the phloem of C. bignonioides cuttings during rooting, IAA: Indole-3-acetic acid; T1: IAA 500 mg·L^{-1} 1 h; T2: IAA 1000 mg·L^{-1} 1 h; T3: IAA 1500 mg·L^{-1} 1 h; T4: IBA 500 mg·L^{-1} 1 h; T5: IBA 1000 mg·L^{-1} 1 h; T6: IBA 1500 mg·L^{-1} 1 h; T7: ABT-1 500 mg·L^{-1} 1 h; T8: ABT-1 1000 mg·L^{-1} 1 h; T9: ABT-1 1500 mg·L^{-1} 1 h; CK: Control.

The rooting type and time points of the cuttings were analyzed. At the callus stage, the IAA content gradually increased. During the formation stage of adventitious roots, the content of IAA decreased. The analysis of the rooting effect of cuttings revealed obvious differences in IAA content among the treatments. Treatments in which the cuttings presented a high rooting rate also had a high IAA content. Among them, the cuttings under the CK had the lowest IAA content and exhibited the lowest root production. Taken together, these analysis results show that the higher the IAA content is, the better the rooting effect.

3.4.2. Changes in GA_3 Content in the Phloem of *C. bignonioides* Cuttings

As shown in Figure 4B, the GA_3 content in the phloem of *C. bignonioides* cuttings during rooting showed a trend of "down–up–down" with time. On days 0–7, the GA_3 content in all the treatments decreased sharply, with treatments 5, 6, and 7 showing the greatest decrease. On days 7–50, the GA_3 contents in treatments 5, 6, and 7 were essentially stable, while those in the other treatments decreased by 50%. On the 15th day, the content of GA_3 showed an increase–decrease trend again, among which the cuttings in the CK treatment presented the largest change. During the whole rooting process, the GA_3 content in the CK treatment was always the highest. Among the various contents, the GA_3 contents in treatments 5, 6, and 7 were maintained at lower levels.

The root morphology and root timing of the cuttings were analyzed. At the callus stage, the GA_3 content decreased. However, the GA_3 content increased during the root primordium and adventitious root generation stage. During the adventitious root elongation stage, the GA_3 content decreased. According to the rooting analysis, the different growth regulators have obvious effects on the GA_3 content and the range of variation within the cuttings. The lower the GA_3 content, the stronger the rooting effect.

3.4.3. Changes in Content of ABA in the Phloem of *C. bignonioides* Cuttings

Figure 4C shows that the ABA content of the phloem of *C. bignonioides* cuttings under each treatment during the rooting process showed a trend of "rising–falling–rising". On days 0–7, the ABA content increased. Among the treatments, the ABA content in the CK increased the most, from the initial value of 117.19 ng·g FW to 218.51 ng·g FW, with an increase of 46.37%. On days 7–30, the ABA content decreased, but on days 30–50, the ABA content increased. On days 0–30, the ABA contents of all treatments were lower than those of the CK. Among the treatments, treatment 5 presented the lowest ABA content.

The rooting morphology and timing of the cuttings were analyzed. At the early callus stage, the ABA content increased. With the production of calli, root primordia, and adventitious roots, the ABA content decreased gradually. During adventitious root elongation, the ABA content began to increase again. Through the analysis of the effects of different growth regulators, concentrations, and treatment times on the cuttings, the lower the ABA content was, the stronger the rooting effect.

3.4.4. Changes in the ZR Content of the Phloem of Cuttings of *C. bignonioides*

Figure 4D shows that the ZR content of the phloem of *C. bignonioides* cuttings under each treatment showed a trend of first decreasing and then increasing with rooting time. On days 0–30, the ZR content increased, but on days 30–50, the ZR content decreased. Among the treatment, treatment 5 presented the largest decrease in ZR content: at the initial cutting stage, the content was 20.59 $ng \cdot g^{-1}$ FW, and the lowest value was 1.93 $ng \cdot g^{-1}$ FW on the 30th day—a decrease of 90.63%. The decrease in ZR content in the CK was the smallest, from the initial value to 9.45 $ng \cdot g^{-1}$ FW on the 30th day—a decrease of 54.10%. Moreover, the ZR content decreased the sharpest in the CK treatment.

The rooting morphology and timing of the cuttings were analyzed. At the callus stage, the ZR content dropped sharply. During the root primordium generation and adventitious root formation stages, the ZR content continuously decreased, with a relatively small rate of decrease. During adventitious root elongation, the ZR content began to increase. Through the analysis of the effects of the different growth regulators, concentrations, and treatment times on the cuttings, the lower the ZR content was, the stronger the rooting effect.

3.4.5. Changes in the IAA/ABA Value of the Phloem of *C. bignonioides* Cuttings

Figure 5A shows that the IAA/ABA value of the phloem of the *C. bignonioides* cuttings under each treatment during the rooting process showed a trend of first increasing and then decreasing. On days 0–30, the IAA/ABA value increased, but on days 30–50, the IAA/ABA value decreased. (The IAA/ABA value for treatment 8 increased on days 0–15 and decreased on days 15–50.) The IAA/ABA values of treatment 5 and treatment 6 increased more than those of the other treatments did. The IAA/ABA value under the CK was always the lowest during rooting. The IAA/ABA values of treatment 5 and treatment 6 peaked on day 30 (1.21 and 1.36), which were 2.81 and 3.16 times the peak values in the CK (0.43), respectively.

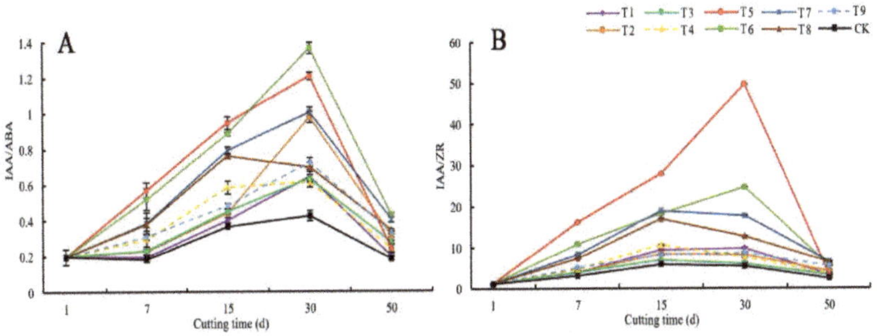

Figure 5. Changes in the IAA/ABA and IAA/ZR value of the phloem of *C. bignonioides* cuttings during rooting. (**A**) Changes in the IAA/ABA value of the phloem of C. bignonioides cuttings during rooting, (**B**) Changes in the IAA/ZR value of the phloem of C. bignonioides cuttings during rooting, IAA/ZR: Indole-3-acetic acid /Zeatin nucleoside; T1: IAA 500 mg·L^{-1} 1 h; T2: IAA 1000 mg·L^{-1} 1 h; T3: IAA 1500 mg·L^{-1} 1 h; T4: IBA 500 mg·L^{-1} 1 h; T5: IBA 1000 mg·L^{-1} 1 h; T6: IBA 1500 mg·L^{-1} 1 h; T7: ABT-1 500 mg·L^{-1} 1h; T8: ABT-1 1000 mg·L^{-1} 1 h; T9: ABT-1 1500 mg·L^{-1} 1 h; CK: Control.

The rooting morphology and timing of the cuttings were analyzed. The IAA/ABA values peaked from the initial period until the formation of adventitious roots. During the adventitious root elongation stage, the IAA/ABA values decreased sharply. Through the analysis of the effects of the different growth regulators, concentrations, and treatment times on the cuttings, the exogenous hormones promoted an increase in the IAA/ABA value. The higher the IAA/ABA value was, the stronger the rooting effect.

3.4.6. Changes in the IAA/ZR Value of the Phloem of *C. bignonioides* Cuttings

Figure 5B shows that the IAA/ZR value of the phloem of the *C. bignonioides* cuttings under each treatment during the rooting process showed a trend of first increasing and then decreasing. On days 0–15, the IAA/ZR value increased, but on days 15–50, the IAA/ZR value decreased. (The IAA/ZR value for treatment 6 increased on days 0–30 and then decreased on days 30–50). Among them, treatment 5 presented the largest increase, from an initial value of 1.13 to a peak value of 49.65 on day 30, equal to an increase of 97.72%. The IAA/ZR value throughout the rooting process was the greatest in treatment 5. The IAA/ZR values of each treatment were higher than those of the CK.

The rooting morphology and timing of the cuttings were analyzed. The IAA/ZR value peaked from the initial period until the formation of adventitious roots. During the adventitious root elongation stage, the IAA/ZR value decreased. The lower the value was, the smaller the decrease. Through the analysis of the effects of the different growth regulators, concentrations, and treatment times on the cuttings, the exogenous hormones promoted an increase in the IAA/ZR value. The higher the IAA/ZR value was, the better the rooting effect.

4. Discussion

4.1. Effects of Growth Regulators on the Rooting Indexes of C. bignonioides Cuttings

According to related studies, rooting types of *C. bignonioides* softwood cuttings include the bark, callus, and mixed rooting types [16,36]. An abundance of research has shown that IBA treatment can significantly increase the rooting of cuttings [37,38]. However, some studies suggest that the rooting effect of cuttings is the greatest under treatment with ABT-1 [39]. Because of the ubiquitous endogenous IAA content plants, root development is always affected by changes in IAA contents. Numerous studies have shown that IAA can promote the rooting of cuttings *of Dalbergia serrata* [40], *Cinnamomum coninna* [41], and *Carnation* [42]. This study showed that three types of growth regulators can promote the formation of adventitious roots of *C. bignonioides* softwood cuttings. In particular, after IBA treatment, calli produced significantly earlier roots, and the rooting rate increased by 76.00% compared with that of the CK. In addition, the number of roots, root length, and rooting index were significantly higher than those in the CK group. Therefore, overall, the rooting effect of IBA treatment was the best, which is consistent with the research conclusions of most scholars of related studies [37,38].

4.2. Changes in the Contents of Endogenous Hormones Associated with Rooting

Plant endogenous hormones are important factors in the formation of adventitious roots on cuttings. Endogenous hormone levels are closely related to the rooting ability of cuttings [43]. IAA is the main hormone that promotes cuttings to develop adventitious roots [44]. According to most related studies [45,46] an increase in IAA content is beneficial to the formation and differentiation of root primordia. According to Taramino et al. high concentrations of IAA can increase the division of root meristem cells, while low concentrations of IAA can accelerate the differentiation of root elongation zone cells [47]. In rooting experiments on cuttings, such as *Chukrasia tabularis* and *Thuja occidentalis,* the content of IAA was found to be greatly increased during the critical period of adventitious root formation and decreased after adventitious root growth [28,48] In the present experiment, hormone treatments promoted IAA synthesis in cuttings, which induced adventitious roots. At this time, calli form in large numbers. Therefore, overall, an increase in IAA content can promote the differentiation of calli to form root primordia and induce rooting, which is consistent with the conclusion of Ma Zhenhua et al. [49] with respect to tetraploid *Robinia pseudoacacia*. Compared with the control group, the cuttings in the IAA-treated group rapidly accumulated, which was favorable for the induction and differentiation of root primordia. The main reason for the early rooting of cuttings may be the early and rapid increase in the IAA content. Therefore, IAA, IBA, and ABT-1 all promoted the IAA content in cuttings. In this experiment, IBA had the best effect and T5 had the best effect on IAA content.

Gibberellin GA_3 mainly plays a role in promoting cell division and elongation and inhibits the formation of adventitious roots [50]. In a study on *Ipomoea fistulosa,* Nanda [51] et al. found that GA_3 can significantly promote the rooting of cuttings and induce leaf bud germination. According to Li Yongxin et al. [33] high concentrations of GA_3 inhibit the formation of adventitious roots. However, low levels of GA_3 promote adventitious roots. In this experiment, the changes in the GA_3 content in cuttings in the treatment group indicated that the formation and elongation of cutting adventitious roots was maintained at a lower level of GA_3. The GA_3 content remained at the lowest level in the IBA-treated group (T5), followed by the IBA-treated group (T6) and the ABT-1-treated group (T7). The GA_3 content of the cuttings of the CK was maintained at a high concentration, and the rooting ability was poor, which also confirms the view that a high concentration of GA_3 inhibits rooting [52]. Therefore, IAA, IBA, and ABT-1 treatment groups promoted the formation and elongation of cutting adventitious root by inhibiting GA_3 content. In this experiment, IBA had the best effect.

ABA is considered a natural hormone in plants and has an inhibitory effect on the rooting of cuttings. The endogenous ABA content of cuttings of difficult-to-root tree species is much higher than that of easy-to-root species [53]. However, some studies [54] have suggested that high concentrations of ABA can promote the rooting of cuttings, the initiation of root primordia, and the formation and development of adventitious roots [55]. According to the results of this experiment, the ABA content in the cuttings showed a trend of increasing and decreasing at the early stage of rooting and then slightly rebounding, which may occur to increase the resistance of the cuttings and reduce the damage caused by stress. During the root primordium induction stage, the ABA content in the cuttings decreased sharply. After adventitious roots developed, the ABA content reached a minimum. Therefore, growth regulators affect the synthesis of ABA, and low levels of ABA are beneficial to rooting. The IBA treatment with the best rooting quality caused the ABA content of cuttings to be maintained at the lowest level, followed by ABT-1 treatment and IAA treatment. The ABA content of the cuttings in the CK treatment changed steadily throughout the rooting process and remained at a high level. In the absence of hormones, the change in the ABA content was small. Moreover, there was a negative correlation between the ABA content of each treatment group and the CK and the rooting ability. Therefore, IBA, IAA, and ABT-1 treatment groups inhibited the production of ABA in cuttings by regulating the changes in endogenous hormones in cuttings and promoted the formation of cutting adventitious roots. In this experiment, the IBA treatment group (T5 and T6) had the best effect.

ZR is the major transport form of cytokinin in woody plants; it plays a role in promoting cell division and differentiation, as well as in the formation of adventitious roots [56,57]. However, many researchers [57,58] have reported that the lower the ZR content, the better the rooting of plant cuttings, especially the induction of root primordia and the production of adventitious roots. The results of this study show that ZR content decreased sharply in the early callus period, which may promote callus induction. During the late callus stage, root primordium stage, and adventitious root generation stage, the ZR content decreased slowly, creating a low-ZR environment for root primordium induction and adventitious root production. After the adventitious root was generated, the ZR content gradually increased, and the elongation of the adventitious root was promoted, which was consistent with findings in *Soapberry* [59]. During the rooting process, the ZR content of cuttings in the treatment group was always lower than that in the control group. Exogenous hormone treatment reduced the ZR content of the cutting base to a certain extent and promoted the hydrolysis of starch and protein, as well as the generation of adventitious roots. Therefore, the IBA, IAA, and ABT-1 treatment groups inhibited the production of ZR in cuttings by regulating the endogenous hormone changes in cuttings and promoted the formation of cutting adventitious roots. In this experiment, the IBA treatment group (T5 and T6) had the best effect.

The rooting of cuttings is the result of the interaction of multiple plant hormones. Changes in the ratios of endogenous hormone contents can reflect the propensity for adventitious root formation [60]. According to studies by Guo Sujuan et al. [61] and Ao Hong et al. [56] high IAA/ABA and IAA/ZR values are beneficial to the rooting of cuttings, as well as the induction and differentiation of root primordia. According to the test results, the ratios of IAA/ABA and IAA/ZR in treatment group T5 cuttings with the best rooting effect all reached their peak on the 30th day after cutting. Therefore, the combined action of exogenous growth regulator influence and IAA, ABA and ZR promoted the formation of calli, the induction of root primordia, and the generation of adventitious roots. Compared with the CK group, the IAA/ABA and IAA/ZR ratios of cuttings in the IAA, IBA, and ABT-1 treatment groups peaked, which was favorable for the induction of root primordia. The main reason for the early rooting of cuttings was the early and rapid increase in the ratios of IAA/ABA and IAA/ZR. IAA, IBA, and ABT-1 treatment all promoted the increase in the IAA/ABA and IAA/ZR ratios. In this experiment, the IBA treatment group (T5 and T6) had the best effect. The results of this experiment support the research conclusions of Guo Sujuan et al. and Ao Hong et al.

5. Conclusions

In conclusion, the endogenous hormone content of cuttings changes in response to plant hormone treatments, thus affecting the rooting of the cuttings. During adventitious root induction, high levels of IAA and ZR and high IAA/ABA and IAA/ZR values favor root primordium induction in *C. bignonioides* softwood cuttings. The reduction in ABA and GA_3 contents is beneficial to the rooting of cuttings. Therefore, IAA and ZR mainly promote *C. bignonioides* softwood cuttings, while ABA and GA_3 play an inhibitory role. IBA treatment increased the IAA content during adventitious root induction and hastened the peak IAA content; however, IBA significantly decreased the ABA and GA_3 contents. This is an important reason for the significant improvement in rooting ability after IBA treatment. The effects of different exogenous hormones on the dynamic physiological and biochemical processes and endogenous hormone contents and the specific mechanism through which endogenous hormones affect the production of adventitious roots of cuttings are complex and need to be further explored.

Author Contributions: Conceptualization, J.Q. and H.B.; methodology, Y.W.; software, Y.W.; validation, R.N. and J.S.; formal analysis, M.M.; investigation, J.Q.; data curation, J.Q.; writing-original draft preparation, J.Q.; writing-review and editing, H.B.; supervision, Y.W.; project administration, H.B.; funding acquisition, H.B. All authors have read and agreed to the published version of the manuscript.

Funding: This research was funded by National Natural Science Foundation of China, grant number 31700549 and Project 948 of State Forestry Administration, grant number 2015-4-04.

Institutional Review Board Statement: Not applicable.

Informed Consent Statement: Not applicable.

Data Availability Statement: All data generated or analyzed during this study are included in this published article.

Conflicts of Interest: The authors declare no conflict of interest.

References

1. Wang, W.C. The Editorial board of "Flora of China" of the Chinese Academy of Sciences was established. *Chin. Sci. Bull.* **1959**, *22*, 772.
2. Zhu, G.F.; Jiang, W.B.; Wen, M.L.; Han, J. Cultural and Historical Connotations of *Catalpa ovata* and Its Ways of Application in Landscape Greening. *Chin. Agric. Sci. Bull.* **2010**, *26*, 301–305.
3. Wei, Z.P. *Evaluation, Screening and Propagation Characteristics of Garden Plants in Hohhot*; Inner Mongolia Agricultural University: Hohhot, China, 2008.
4. Bi, H.T.; Wang, Y.G.; Shen, J.M.; Quan, J.E. Effects of IBA treatment on rooting of cutting and physiological and biochemical twig of *Catalpa bignonioides*. *J. Henan Agric. Univ.* **2020**, *54*, 44–52.
5. Chen, S.C.; Wang, X.J.; Xiao, Z.D.; Sun, Q.; Xiao, D.J. Primary Report on the Propagation Experiment of *Catalpa bungei* by Twig Cutting. *J. Anhui Agric. Sci.* **2008**, *36*, 7635-7636, 7901.
6. Liang, Y.W.; Peng, F.R.; Wang, S.C. Preliminary report on cutting experiment of *Catalpa bungeana* twig. *J. For. Eng. J.* 2006.
7. Zhao, X.Y.; Wang, J.H.; Zhang, J.F.; Zhang, S.F.; Zhang, J.G.; Ma, J.W. Study on phenotypic traits and germination characters of four taxons of catalpa genus seed. *J. Northwest AF Univ. (Nat. Sci. Ed.)* **2008**, *12*, 31.
8. Peng, C.; Li, Z.F.; Xu, X.H.; Huang, G.W.; Chen, H.L. Effect of Different Phytohormone and Their Concentraions and Presoaking Time on Seeds Germination of *Catalpa ovata* Seed. *Seed* **2016**, 35.
9. Cui, L.J.; Chen, S.L.; Liu, Y.X.; Lin, J.; Shi, K.M. Germination response of catalpa bungei seed to exogenous hormones treatment under different temperatures. *J. Northeast For. Univ.* **2020**, *2*, 12–16.
10. Hu, T.; Cao, Y.; Zhang, G.X. Softwood Cutting of *Chionanthus virginicus* and Changes of Endogenous Hormone Content during Rooting Period. *J. Northwest For. Univ.* **2019**, *34*, 115–121.
11. Ren, X.L. Research on cutting techniques and cuttings rooting type in American *Nandina domestica'Alba'*. *Non-Wood For. Res.* **2019**, *37*, 188–192.
12. Zhu, Y.C.; Li, B.; Liao, W.B. Effects of three auxins treatments on cutting root ting of *Lonicera korolkowi 'Zabclii'*. *Pratacult. Sci.* **2016**, *33*, 61–66.
13. Tong, Z.Q.; Wei, J. Cuttage Propagation Technique and Rooting Me chanism of *Quercus mongolica* with Softwood Cuttinng. *J. Northwest For. Univ.* **2017**, *32*, 116–121.
14. Bai, L.; Li, R.S.; Yi, G.T.; Yang, J.; Zou, W. Rooting factors and optimization for propagation of *Mytilaria laosensis* cuttings. *J. Zhejiang AF Univ.* **2016**, *33*, 543–550.
15. Quan, J.E.; Meng, S.; Guo, E.H.; Zhang, S.; Zhao, Z.; Yang, X. De novo sequencing and comparative transcriptome analysis of adventitious root development induced by exogenous indole-3-butyric acid in cuttings of tetraploid *black locust*. *BMC Genom.* **2017**, *18*, 179. [CrossRef] [PubMed]
16. Wang, X.L.; Zhao, Z.; Quan, J.E.; Zhang, X.; Zhang, B. Rooting and Correlative Enzyme Activities of Hardwood Cuttings of Tetraploid *Robinia pseudoacacia*. *Acta Bot. Boreali-Occident. Sin.* **2011**, *31*, 116–122.
17. Wang, Q.; Zhang, J.; Zhong, C.L.; Zhang, Y.; Wei, Y.C.; Meng, J.X. Variation of endogenesis hormone and nutritive matter concentration in *Chukrasia tabularis* cuttings during rooting. *J. Cent. South Univ. For. Technol.* **2020**, *40*, 111–119.
18. Zhai, Y.F.; Liu, X.D.; Lv, D.; Zhao, M.; Zhao, H.; Zhao, X.P.; Wang, Y.L.; Zhao, Y.H. Effects of plant growth regulators on the rooting and enzyme activities changes of Lonicera tatarica cuttings. *J. Cent. South Univ. For. Technol.* **2021**, *41*, 52–61.

19. Wang, X.L. *Physiological and Biochemical Basis of Adventitious Roots in Tetraploid Locust Cutting*; Northwest A&F. University: Xi'an, China, 2011.
20. Zhang, E.L.; Ma, L.L.; Yang, R.T.; Li, L.; Wang, Q.; Li, Y.; Wang, P. Transcriptome Profiling of IBA-Induced Adventitious Root Formation in Softwood Cuttings of *Catalpa bungei'Yu-1'*4. *Sci. Silvae Sin.* **2018**, *54*, 48–61.
21. Ma, L.L.; Wang, P.; Zhang, Z.Y.; Li, L.F.; Yang, R.T.; Li, Y. Evaluation on Rooting Ability of Softwood Cuttings on *Catalpa*. *North. Hortic.* **2014**, *15*, 72–77.
22. Wang, G.P.; Wang, L.G.; Wang, X.C.; Zhang, C.; Zhang, L.; Liu, B. Dynamic characteristics of cutting rooting of *Catalpa bungei* with tender branches. *J. Nanjing For. Univ.* **2020**, *44*, 94–102.
23. Zhen, J.H.; Dong, Q.; Duan, H.C. Quality evaluation on *Crateva unilocularis* cuttings treated with three hormones. *Non-Wood For. Res.* **2020**, *38*, 230–237.
24. Wang, X.S.; Bao, Z.L.; Chen, L.; Cui, L.; Li, F.; Zhang, Y.B. Green Cutting Propagation of Cold Resistant Prunus salicina Linn. *J. Northeast For. Univ.* **2021**, *49*, 13–16.
25. Liang, X.C.; Huang, S.D.; Jiang, Y.M.; Zhao, C.M. Effects of Exogenous Hormones and Substrate on Cutting Rooting of *Cryptocarya concinna*. *Chin. Agric. Sci. Bull.* **2021**, *37*, 24–29.
26. Yan, Z.J.; Li, H.M.; Yang, P.; Gao, W.S.; Jian, H.Y. Effects of Plant Growth Regulators on Cutting Propagation of *Rosa odorata var. Odorata*. *Acta Agric. Jiangxi* **2020**, *32*, 41–44.
27. Wu, K.Y.; Gong, B.C.; Xu, Y.; Zhang, P. Effects of different plant growth regulators on hardwood cuttage of *persimmon rootstock Yalin* 6. *Acta Agric. Zhejiangensis* **2021**, *33*, 1256–1263.
28. Liu, M.G.; Wang, L.; Dong, S.J.; Cai, X. Endogenous hormone variation in cuttings of *Thuja occidentalis* L. in the period of adventitious root formation. *J. Shenyang Agric. Univ.* **2010**, *41*, 555–559.
29. He, C.D.; Yao, R.; Huang, R.C.; Mi, X.Q.; Zhang, C.; Tan, Y.M.; Wang, Y.Q. Effects of K-IBA treatments on adventitious rooting and endogenous hormones contents of shoot cuttings of *Thuja occidentals* L. *J. Cent. South Univ. For. Technol.* **2017**, *37*, 7–12.
30. Zhang, X.P.; Fang, Y.M.; Huang, S.H. The endogenous hormone's variation during theadventitious roots formation of *hybrid Tuliptrees* by cutting. *J. Nanjing For. Univ.* **2004**, *28*, 79–82.
31. Li, C.C.; Zhao, Y.L.; Zhang, D.L.; Wu, H.; Chen, X. Changes of endogenous hormones and anatomical structure of *Rhododendron stamineum* during cutting propagation. *For. Res.* **2012**, *25*, 360–365.
32. Zhao, Y.L.; Chen, X.; Li, C.C. Dynamic of physiology and biochemistry duringwild *Rhododendron scabrifolium* cutting propagation. *Sci. Silvae Sin.* **2013**, *49*, 45–51.
33. Li, Y.X.; Zeng, H.J.; Wang, X.M.; Cai, N. Changes of endogenous hormones during *Swida wilsoniana* wanger cutting. *Chin. Agric. Sci. Bull.* **2010**, *26*, 247–251.
34. Dong, S.J.; Liu, M.G.; Dai, F.; Wu, Y.L.; Shan, S.T.; Ding, R.J. Variation of endogenous hormone contents in softwood cuttings of Armeniaca *sibirica* during adventitious root formation. *Non-Wood For. Res.* **2013**, *31*, 108–114.
35. Shi, F.H.; Zhao, R.; Luo, S.; Zhen, C.; Shen, Y.B. Study on plant hormone changes in soft cuttings of *Tiliamiqueliana* during rooting. *J. Cent. South Univ. For. Technol.* **2019**, *39*, 21–26.
36. Sun, A.; Li, L.F.; Guo, L.; Su, N. Reponse of Rooting Traits of *Machilus longipedicellata* Rooted Cutting to IBA, NAA and IAA. *North. Hortic.* **2017**, *16*, 102–106.
37. Wiesman, Z.; Lavee, S. Enhancement of IBA stimulatory effect on rooting of *olive* cultivar stem cuttings. *Sci. Hortic.* **1995**, *62*, 189–198. [CrossRef]
38. Copes, D.L.; Mandel, N.L. Effects of IBA and NAA treatments on rooting *Douglas-fir stemcuttigs*. *New For.* **2000**, *20*, 249–257. [CrossRef]
39. Wang, X.M.; Peng, D.Q.; Wu, W.L.; Zhu, H.; Huang, T.; Lv, L.F.; Li, W.L. Effects of culture Medium and plant growth regulators on rooting of softwood cuttings of *beach plum*. *J. Northwest For. Univ.* **2014**, *29*, 114–118.
40. Xu, S.S.; Liu, X.J.; Xu, D.P.; Hong, Z.; Guo, J.-Y.; Yang, Z.-J. Influence of IAA and NAA on cutting propagation of *dalbergia odorifera*. *For. Res.* **2021**, *34*, 168–176.
41. Zheng, J.H.; Dong, Q.; Duan, H.C.; Cha, X.F.; Zhou, C.; Zhou, Z.C. Effect of Three Growth Regulators on Rooting of Cuting of Crateva unilocularis. *Bull. Bot Res.* **2020**, *40*, 202–208.
42. Garrido, G.; Guerrero, J.R.; Cano, E.M. Origin and basipetal transport of the IAA responsible for *carnation* cuttings. *Physiol. Plant.* **2002**, *31*, 302–312. [CrossRef]
43. Haussig, B.E. Influence of indole-3-acetic acid on adventitious root primordia of *brittle willow*. *Planta* **1970**, *95*, 27–35. [CrossRef] [PubMed]
44. Trefois, R.; Brunner, T. Influence of the endogenous auxinic contene on the response to propagation by cuttings and on the responce to propagation by cuttings and on the dwarfing effect of some types of *Prunus*. *Bot.-Bot. Publ.* **1982**, *69*, 197–204.
45. Xu, J.Z.; Chen, S.W. The effect of the endogenous hormone's contents (ABA and IAA) in hardwood cuttings of peach to rooting. *Acta Hortic. Sin.* **1989**, *4*, 275–278.
46. Lv, M.; Fang, Y.M.; Jia, H.; Wang, A.X. Dynamics of endogenous hormones during the process of adventitious root formation in *Red Alder*. *Dev. For. Sci. Technol.* **2009**, *23*, 16–19.
47. Taramino, G.; Sauer, M.; Stauffer, J.L., Jr.; Multani, D.; Niu, X.; Sakai, H.; Hochholdinger, F. The maize (*Zea mays* L.) RTCS gene encodes a LOB domain protein that is a key regulator of embryonic seminal and post-embryonic shoot-borne root initiation. *Plant J.* **2007**, *50*, 649–659. [CrossRef]

48. Wang, Q. *Propagation Techniques and Rooting Mechanism of Chukrasia Tabularis Softwood Cuttings*; Northeast Forestry University: Harbin, China, 2019.
49. Zhao, X.M.; Huo, C.F.; Shen, H.L. Studied on changes hormone in the periods of making roots in the softwood of *Robinia pseudoacacia*. *J. Northwest For. Univ.* **2013**, *28*, 18–23.
50. Han, H.; Zhang, S.G.; Sun, X.M. A review on the molecular mechanism of plants rooting modulated by auxin. *Afr. J. Biotechnol.* **2009**, *8*, 348–353.
51. Nanda, K.K.; Anand, V.K.; Chibbar, R.N. The promotive Effect of gibberellic on the production of adventitious roots on stem cuttings of *Ipomoea fistulosa*. *Planta* **1972**, *105*, 360. [CrossRef]
52. Sun, X.D.; Zhang, P.; Shen, S.L. Effects of Rooting Agents and Cutting Treatment on Rooting of Shoot Cuttings of *Acer mono*. *J. Northest. For. Univ.* **2018**, *46*, 5–8.
53. Chi, Y.; Hou, Y.L.; Yu, Y.J.; Yu, D.Y.; Feng, B.M.; Yao, Z.A.; Jiang, G.B. Dynamic Changes of Endogenous Hormones Contents in Mutant of *Populus tomentosa* and its Relative Species in the Cutting and Rooting Procedure. *J. Anhui Agric. Sci.* **2007**, *35*, 8438–8439.
54. Lin, S.J.; Jiang, J.; Feng, X.; Li, T.H.; Chang, W.X. The Study of Adventitious Root Generation in Leaves of Tissue Culture of (*P.simonii* × *P.nigra*) × P.15 ACL in the Relation to Exogenous Hormones and Endogenous Hormones. *For. Sci. Technol.* **2016**, *21*, 68–79.
55. Huang, Z.L.; He, H.K.; Cao, Y.Y.; Peng, Y.H.; Wang, X.J.; Liu, X. Changes of endogenous hormones in Castanopsis hystrix cuttings. *J. Cent. South Univ. For. Technol.* **2015**, *2*, 22–25.
56. Ao, H.; Wang, K.; Feng, Y.L. Endogenous hormones levels in cuttings of *Larix olgensis* and the irrelations to rooting. *Bull. Bot. Res.* **2002**, *22*, 190–195.
57. Ao, H.; Wang, K.; Feng, Y.L. Qualitative Analysis of Endogenesis Hormone and Polyphenol During Rooting of Cuttings in Norway Spruce (*Picea abies*). *For. Sci.* **2015**, *51*, 155–162.
58. Han, J.H.; Shao, W.H.; Liu, J.F.; Diao, S.F. Content changes of endogenous hormone and polyphenols during hardwood cuttage progress in *Sapindus ukorossi*. *Non-Wood For. Res.* **2019**, *37*, 37–43+51.
59. Qiao, Z.Q.; Wang, X.M.; Zeng, H.J.; Li, Y.X.; Cai, N.; Wang, X.Y. Changes of endogenous hormones during *Lagerstroemia indica* 'Xiang Yun'cutting. *Hunan For. Sci Technol.* **2015**, *42*, 49–53.
60. Zhou, Z.Z.; Liu, S.C.; Zhang, J.H.; Liang, K.N.; Ma, H.M.; Huang, G.H. Effects of IBA treatments on adventitious rooting and endogenous hormones contents of shoot cuttings of *Callicarpa nudiflora*. *Chin. J. Trop. Crops* **2016**, *37*, 1075–1080.
61. Guo, S.J.; Lin, H.Q.; Li, F.L. Progress of Study on Rooting Anatomy and Physiology of Forest Tree Cuttings. *J. Beijing For. Univ.* **1997**, *4*, 64–69.

life

MDPI

Article

Identification and Functional Analysis of SabHLHs in *Santalum album* L.

Ting Zhang [1,2], Xiaohong Chen [1,2], Yuping Xiong [1], Meiyun Niu [1,2], Yueya Zhang [1,2], Haifeng Yan [3], Yuan Li [1], Xinhua Zhang [1] and Guohua Ma [1,*]

1 Guangdong Provincial Key Laboratory of Applied Botany, South China Botanical Garden, The Chinese Academy of Sciences, Guangzhou 510650, China; zhangting13jmk@163.com (T.Z.); chenxiaohong@scbg.ac.cn (X.C.); xiongyuping@scbg.ac.cn (Y.X.); niumeiyun@scbg.ac.cn (M.N.); tulipazhyy@163.com (Y.Z.); liy@scbg.ac.cn (Y.L.); xhzhang@scbg.ac.cn (X.Z.)

2 University of the Chinese Academy of Sciences, Beijing 100049, China

3 Cash Crop Institute of Guangxi Academy of Agricultural Sciences, Nanning 530007, China; gstsyhf@163.com

* Correspondence: magh@scbg.ac.cn; Tel.: +86-020-3735-2993

Abstract: *Santalum album* L., a semi-parasitic evergreen tree, contains economically important essential oil, rich in sesquiterpenoids, such as (Z) α- and (Z) β-santalol. However, their transcriptional regulations are not clear. Several studies of other plants have shown that basic-helix-loop-helix (bHLH) transcription factors (TFs) were involved in participating in the biosynthesis of sesquiterpene synthase genes. Herein, bHLH TF genes with similar expression patterns and high expression levels were screened by co-expression analysis, and their full-length ORFs were obtained. These bHLH TFs were named *SaMYC1*, *SaMYC3*, *SaMYC4*, *SaMYC5*, *SabHLH1*, *SabHLH2*, *SabHLH3*, and *SabHLH4*. All eight TFs had highly conserved bHLH domains and *SaMYC1*, *SaMYC3*, *SaMYC4*, and *SaMYC5*, also had highly conserved MYC domains. It was indicated that the eight genes belonged to six subfamilies of the bHLH TF family. Among them, SaMYC1 was found in both the nucleus and the cytoplasm, while SaMYC4 was only localized in the cytoplasm and the remaining six TFs were localized in nucleus. In a yeast one-hybrid experiment, we constructed decoy vectors pAbAi-SSy1G-box, pAbAi-CYP2G-box, pAbAi-CYP3G-box, and pAbAi-CYP4G-box, which had been transformed into yeast. We also constructed pGADT7-*SaMYC1* and pGADT7-*SabHLH1* capture vectors and transformed them into bait strains. Our results showed that *SaMYC1* could bind to the G-box of *SaSSy*, and the *SaCYP736A167* promoter, which *SaSSy* proved has acted as a key enzyme in the synthesis of santalol sesquiterpenes and *SaCYP450* catalyzed the ligation of santalol sesquiterpenes into terpene. We have also constructed pGreenII 62-SK-*SaMYC1*, pGreenII 0800-LUC-*SaSSy* and pGreenII 0800-*LUC-SaCYP736A167* via dual-luciferase fusion expression vectors and transformed them into *Nicotiana benthamiana* using an *Agrobacterium*-mediated method. The results showed that *SaMYC1* was successfully combined with *SaSSy* or *SaCYP736A167* promoter and the *LUC/REN* value was 1.85- or 1.55-fold higher, respectively, than that of the control group. Therefore, we inferred that *SaMYC1* could activate both *SaSSy* and *SaCYP736A167* promoters.

Keywords: bHLH transcription factor; dual luciferase; gene cloning; sandalwood; *SaSSy*; *SaCYP736A167*; subcellular localization; yeast one-hybridization; dual luciferase activity

Citation: Zhang, T.; Chen, X.; Xiong, Y.; Niu, M.; Zhang, Y.; Yan, H.; Li, Y.; Zhang, X.; Ma, G. Identification and Functional Analysis of SabHLHs in *Santalum album* L. *Life* **2022**, *12*, 1017. https://doi.org/10.3390/life12071017

Academic Editors: Sen Meng and Jie Luo

Received: 13 May 2022
Accepted: 24 June 2022
Published: 8 July 2022

1. Introduction

Transcription factors (TFs) are key regulatory elements in plants that often bind to *cis*-acting elements (CAEs) in the promoter region upstream of a gene, and regulate its expression. They typically had four functional regions, a transcriptional regulatory region, a nuclear localization signal region, an oligomerization site region, and a DNA-binding region [1]. The amino-acid sequences of TF DNA-binding region determine its family, such as basic-helix-loop-helix (bHLH), MYB, WRKY, bZIP, MADs, TCP, AP2/ERF, or other families of TFs. Many TFs were involved in plant growth and development, secondary metabolism, stress resistance and other processes, but the TFs involved in regulating synthesis of sesquiterpenes usually fell into four families: AP2/ERF, bHLH, MYB, and WRKY [2–4]. Among them, bHLH TFs were the second largest family of TFs in plants after the MYB TFs, and they had a highly conserved domain that was divided into two regions—an alkaline region located at the N-terminus, which consisted of 15–20 amino acids, and another α helix 1-ring-α helix 2 region, located at the C-terminus, which consisted mainly of hydrophobic amino acids [5]. This domain consisted of about 60 amino acids, 25 of which were conserved residues, five were in the alkaline region, six were in the first spiral region, two were in the ring, and another 12 were in the second spiral region [6]. Alkaline regions could bind to the E-box (5′-CANNTG-3′) and the G-box (5′-CACGTG-3′) in DNA sequences [7]. The α-helix 1-cyclic-α helix 2 region contained many hydrophobic amino acids, and in order to be functional, it often formed homo- or heterodimers [8].

bHLH TFs were widely involved in the growth and development of plants. S1PRE2, a bHLH TF was highly expressed in immature green *Solanum lycopersicum* fruits after induction by gibberellic acid (GA_3), and as the *S1PRE2* gene was silenced, fruits became smaller and pericarps became thinner, indicating that S1PRE2 was a positive regulator during fruit development [9]. bHLH TFs formed complexes with MYB TFs, activated the expression of key genes that regulated stamen development, seed germination, and seedling development in *Arabidopsis thaliana* [10]. bHLH TFs SPEECHLESS (SPCH), MUTE, FAMA, and ICE/SCREAM (SCRM) co-regulated the formation of plant stomata via signal transduction [11,12]. The bHLH-like TF LAX/ba1 co-regulated branching and inflorescence branching with GLAS family members Ls, LAS, and MOC1, and an R2R3-type MYB family member Bl [13].

Signal transduction is a very important process in plants because their response to external stimuli takes place via signal transduction. bHLH TFs played a key negative regulatory role in plant pigment signal transduction. They were also involved in plant hormone signal transduction [14]. Three bHLH TFs (*BEE1*, *BEE2* and *BEE3*) were regulators required for the early response of *A. thaliana* brassinosteroid (BR), as demonstrated by their mutants *bee1*, *bee2*, and *bee3* [15]. In *A. thaliana*, the bHLH TF *AtMYC2* upregulated an abscisic acid (ABA)-inducible gene, while a mutant of *AtMYC2* downregulated an ABA-inducible gene, demonstrating that it acted as a positive regulator in ABA-induced gene expression [16]. *AtMYC2* was also involved in the signal transduction pathway of Jasmonate-ZIM, which acted as a transcriptional inhibitor [17]. In *Malus pumila*, *MdbHLH3* activated the transcription of genes that regulated ethylene biosynthesis (*MdACO1*, *MdACS1*, and *MdACS5A*), thereby promoting the synthesis of ethylene [18]. The bHLH TF *PIF4* played a major role in multiple signal integration during plant growth regulation, serving as a positive regulator in cell elongation, and its activity was regulated by various environmental signals and hormonal signals including GA_3, auxin, and BR, as well as light and temperature, both transcriptionally and post-translationally [19]. In *Oryza sativa*, nuclear localization of the TF *OsbHLH073* was involved in regulation of plant height, and internodal and panicle elongation by downregulating the biosynthesis of GA_3 [20]. High temperatures might increase both epidermal *PIF4* transcription and the epidermal *PIF4* DNA-binding ability in *A. thaliana* [21].

Santalum album L. is a semi-parasitic tree that belongs to the Santalaceae family. It has a high economic value, which is mainly reflected in its heartwood, which is often used as a raw material for carving crafts, and it is often made into incense commonly used in perfume, while sandal essential oil extracted from its heartwood has displayed anti-cancer [22,23], antioxidant [24], anti-inflammatory and analgesic [25,26] properties, and has been used in the treatment of skin diseases [27,28]. The main components of sandal essential oil are α- and β-santalol [29]. Therefore, it is necessary to understand biosynthesis of the main sandal sesquiterpenes.

In recent years, an increasing amount of research has been dedicated to synthesis of sandal sesquiterpenes, which were mainly synthesized by the mevalonic acid pathway [30]. *SaSSy* and its homologous genes *SauSSy* and *SpiSSy* regulated synthesis of terpenoids such as α- and β-santalol, while the strongest regulatory function shown by *SaSSy*, and *SaSSy* acted as a key enzyme in synthesis of sandal sesquiterpenes [30]. SaCYP450 family enzymes catalyzed the ligation of sandal sesquiterpenes into terpene [31]. Among them, *SaCYP736A167* converted α- and β-santalol into (Z)-α- and (Z)-β-santalol [30,32]. The farnesyl diphosphate synthase gene *SaBS* was cloned from *S. album*, and it encoded an enzyme necessary for catalytic synthesis of the substrates, such as (*E*, *E*)-farniki pyrophosphate [33]. Three new terpene synthase (*TPS*) genes, *SaTPS1*, *SaTPS2* and *SaTPS3*, were isolated from *S. album*: while *SaTPS2* and *SaTPS3* catalyzed synthesis of (*E*)-α-bergamotene, (*E*)-β-farnesene and β-bisabolene, *SaTPS1*, *SaTPS2* and *SaTPS3* responded to hormones and abiotic stresses [34]. A *TPS* gene located in chloroplasts and the cytoplasm was isolated from *S. album*, the enzyme encoded by this gene mainly catalyzed synthesis of linalool and nerolol, which were secondary components of sandal essential oil, while this gene responded to abiotic stress [35]. In recent years, many studies have focused on upstream regulatory genes of key enzyme genes in the biosynthetic pathway of sandal oil, such as the *SaAACT* and *SaHMGS* genes, which regulated the synthesis of important substrates, and whose function was verified in yeast by complementation experiments [36,37]. *SaDXR* was a 1-deoxy-D-xylulose-5-phosphate reductoisomerase (DXR), cloned from *S. album*, which played an important role in the biosynthesis of photosynthetic pigments and shifted the flux to sesquiterpenoids [38].

The transcriptional regulation of sesquiterpenes by bHLH TFs in *S. album* has not been reported. The objective of this work was to identify bHLH TF genes in *S. album* transcriptome. The physicochemical properties were determined; bioinformatics and subcellular localization analyses were also performed. To explore whether these bHLH TF genes were involved in the expression of key enzyme genes that regulated the synthesis of santalol, some promoters of key enzyme genes (*SaSSy* and *SaCYP450*) were explored by yeast one-hybridization and dual-luciferase experiments. Our findings will provide a theoretical basis for additional studies of bHLH TFs, to assess their regulation of the synthesis of sandal sesquiterpenes.

2. Materials and Methods

Plant materials: The material used in this experiment included a 10-year-old *S. album* tree, which planted in the sandalwood research base of South China Botanical Garden of the Chinese Academy of Sciences, Guangzhou. Wild-type *Arabidopsis* seeds were preserved and grown in incubators at 22–23 °C in the Lab, 16-h photoperiod, 100 μmol m^{-2} s^{-1}. *Nicotiana benthamiana* was grown at day/night 16-h photoperiod of 28/26 °C, 80 μmol m^{-2} s^{-1}, in an incubator used for subsequent transient expression.

2.1. Reagents

Kits: 1% agarose gel DNA recovery kit, plasmid medium volume kit were purchased from Magen BioTech Co., Ltd. (Guangzhou, China). A dual luciferase activity assay kit was purchased from Promega (Beijing, China). Yeast one-hybrid kits and yeast ligation kits were purchased from Clontech (Terra Bella Avenue Mountain View, CA, USA).

Enzymes: 10× loading buffer, DL 2000 DNA Marker, pMD18-T, rTaq enzyme, T4 ligase, In-Fusion HD enzyme premix and various restriction nucleic acid endonucleases were purchased from TakaRa Bio Inc. (Dalian, China); KOD FX was purchased from OYO TBO (Osaka, Japan); and 2× Flash PCR MasterMix (Dye) were purchased from Kangwei Century Co (Beijing, China).

Culture medium: The components of the LB medium, yeast extract, tryptone, and sodium chloride were purchased from Oxoid Biological Company and Aladdin Biological Company, respectively.

Other reagents: Cellulase Cellulose R10 and pectinase Macerozyme used for subcellular localization were purchased from Yakult Honsha in Japan; Bovine Serum Protein (BSA) was purchased from Sigma (Merck KGaA, Darmstadt, Germany).

Vector and Escherichia coli lines: The subcellular localization vector pSAT6-EYFP-N1 was preserved by our laboratory; the *E. coli* DH5α line was purchased from Shanghai Vidi Biotechnology Co., Ltd. (Shanghai, China). Guangzhou Qingke Biotechnology Co., Ltd. (Guangzhou, China) provided the service for primer synthesis and sequencing.

2.2. Screening and Cloning of bHLH Transcription Factors

Based on our group's existing sandal-tree-transcriptome data and the research achieved [3,35], eight bHLH TFs (SabHLHs) with similar expression patterns (more expression in heartwoods than expression in sapwoods) and similar expression patterns to *SaSSy* and *SaCYP736A167*, which regulated sandal oil biosynthesis, were screened by co-expression analysis. Mixed cDNA of the stems and leaves from the 10 years old sandal tree was used as template. PCR amplifications were carried out with TaKaRa rTaq enzyme and corresponding primers. PCR products were separated by electrophoresis agarose gel (1%) electrophoreses and recovered using the gel recovery kit (Meiji Biotech) according to the instructions. The purified PCR product was ligated overnight with T4 ligase with pMD18-T vector and transformed into *E. coli*. Single colonies containing the fragment of interest were picked and inoculate to liquid LB medium containing Amp antibiotics, incubated (37 °C, 200 rpm) for 12 h. The *E. coli* solution was sequenced in Qingke Biotech.Plasmids were extracted using plasmid small lifting kit (Meiji Biotech) according to the instructions. Eight TF ORF plasmids were then obtained (Supplementary Table S1). The ORF sequences of these eight TFs were submitted to NCBI for the registration numbers (Supplementary Table S2).

2.3. Bioinformatics Analysis of SabHLHs

Based on the sequence of SabHLHs, the amino acid length, molecular weight, isoelectric point, instability coefficient and mean hydrophilicity of these eight TFs were analyzed using the online website ExPASY (https://web.expasy.org/protparam/) (accessed on 21 February 2021). After translated by DNAstar editing, the conservative domain (Motif) of SabHLHs was analyzed using the online website MEME (http://meme-suite.org/tools/meme) (accessed on 21 February 2021) and graphed with TBtools software (https://www.tbtools.com/) (accessed on 21 February 2021) [16]. MEME parameters were set according to our previous work [3]. Protein sequences of bHLH from the pattern plant *A. thaliana* were downloaded from Phytozome (https://phytozome.jgi.doe.gov/pz/portal.html) database (accessed on 26 November 2020); multi-sequence alignment analysis was performed using ClustalX 2.0 (Supplementary Table S3). A systematic evolutionary tree was constructed using the Neighbor-Joining method in MEGA 7.0 [39], where the number of bootstraps was set to 1000.

2.4. Subcellular Localization Analysis

pSAT6-EYFP-N1 plasmid was digested by *Bam*HI and *Eco*RI restriction endonucleases. TF fragments which removed stop codon were amplified using TaKaRa's KOD FX and then cloned into pSAT6-EYFP-N1 vector by homologous recombinant ligase (TaKaRa). The recombinant vectors were transformed into *E. coli*. Positive colones were sequenced and plasmids were extracted.

We bathed 10 mL enzymatic solution for 10 min and cooled to room temperature, then added sterilized 100 μL $CaCl_2$ (1.0 M) and 100 μL Bovine serum albumin (BSA). We then selected *A. thaliana* that was growing well and teared off the epidermis of *Arabidopsis* leaves with scotch tape and put them into a Petri dish containing the above enzymatic solution, and then incubated at 50 rpm at 22 °C for 3 h under weak light conditions. Slowly, we added W5 solution (150 mL NaCl, 125 mL $CaCl_2$, 5 mM KCl, 2 mM ES, osmolarity 550–580 mOsm), pH5.7 (KCl), then stored at 4 °C; the amount of W5 solution was added depending on the number of cells; the final solution color with light green was preferred. We suspended the pellet gently; Microscopic examination to ensure the integrity and concentration of protoplasts was appropriate. No-load pSAT6-EYFP-N1 was used as a control.

2.5. Interaction Detection of Transcription Factors SaMYC1, SabHLH1 and G-Box Elements

Eight transcription factors (*SaMYC1, SaMYC3, SaMYC4, SaMYC5, SabHLH1, SabHLH2, SabHLH3, SabHLH4*) were constructed into pGADT7-AD and transformed to the Y1H Gold strain to express the capture protein. We inoculated the well-grown single colonies into 3 mL YPDA liquid medium and shook to OD 0.2, then diluted with 0.9% NaCl solution for 100 fold to an OD value at 0.002, then took 4.5 μL of dots in SD/-Leu/AbA 0 ng/mL and SD/-Leu/AbA 200 ng/mL media, respectively.Sequences of *SaSSy* and *SaCYP736A167* promoters were submitted to the online software plantCARE (http://bioinformatics.psb.ugent.be/webtools/plantcare/html/) (SaSSy and SaCYP736A167 were both accessed on 11 December 2019) for G-Box and E-Box prediction.

SaSSy promoter contained G-Box upstream and downstream of the small fragment *SSy1G*. *SaCYP736A167* promoter contained G-Box 10bp upstream and downstream of small fragments *CYP1G, CYP2G, CYP3G, CYP4G*, respectively. The synthesis sequences by Qingke Biotech. Com. were shown in the Supplementary Table S4. *pAbAi-SSy1G*-Box, *pAbAi-CYP1G*-box, *pAbAi-CYP2G*-box, *pAbAi-CYP3G*-box and *pAbAi-CYP4G*-box plasmid were integrated into Y1H Gold yeast strain after restriction endonuclease BstbI monoenzyme cleavage. Positive single colony was shaken in 3 mL YPDA liquid medium to OD 0.2, and then adjusted the OD value to 0.002 with 0.9% NaCl solution, and 100 μL was screened for the lowest AbA inhibition concentrations in SD/-Ura media with different AbA concentrations.

AD-SaMYC1 and the negative control were transformed into wild type bait strains and mutant element bait strains. The transformed strain was first cultured on SD/-Leu medium that contained no AbA, after the colonies grew up, a well-grown monoclonal was picked. As the YPDA liquid medium was shaken to a bacterial liquid OD value of 0.2, and diluted with 0.9% NaCl solution to 100-fold to an OD value at 0.002, 4.5 μL of liquid was taken on the medium of SD/-Leu with corresponding concentration of AbA. After this was incubated at 30 °C for 3–5 d, the results were obtained.

2.6. SaMYC1 Activated SaSSy and SaCYP736A167 Promoter Activity

In order to further determine the regulatory effect of *SaMYC1* on *SaSSy* and *SaCYP736A167* genes, the transcription factor *SaMYC1* was constructed on a pGreenII 62-SK vector driven by a 35S promoter as an effective carrier. *SaSSy* and *SaCYP736A167* promoter sequences were constructed on a reporter vector containing *REN* gene and *LUC* gene pGreenII 0800-LUC vector as a reporter vector, which renilla luciferase gene *REN* was initiated by a 35S strong promoter on the vector, so renilla luciferase activity was used as a reference. The firefly luciferase gene was initiated and expressed by the *SaSSy* promoter. The transcription factor vector and promoter vector were transformed into *Agrobacterium* competent cell GV3101, respectively, which the OD value of bacterial fluid was 0.6–0.8, and then transformed into *N. benthamiana* leaves according to the ratio of transcription factor: promoter = 10:1 (*v:v*). After co-incubation for 2 days under normal conditions, the effect of *SaMYC1* on *SaSSy* promoter was determined by detecting the chemiluminescence values of firefly luciferase and renilla luciferase. The ratio of no-load pGreenII 62-SK+ *SaSSy*-0800-LUC, pGreenII 62-SK + *SaCYP736A167*-0800-LUC was used as the control, and the ratio of firefly luciferase chemiluminescence value divided renilla luciferase chemiluminescence value was *LUC/REN*.

2.7. Statistical Analysis

The experiment statistics were analyzed using IBM SPSS 19.0 (IBM Corp., Armonk, NY, USA), and Duncan's multiple range test at $p \leq 0.05$ to denote significant differences between the means. Different letters indicated a significant difference. All data represented three biological replicates of mean ± standard errors (SE).

3. Results

3.1. Cloning SabHLHs ORFs

Eight bHLH TFs were screened and named *SaMYC1, SaMYC3, SaMYC4, SaMYC5, SabHLH1, SabHLH2, SabHLH3,* and *SabHLH4* (Supplementary Tables S1–S3). A heat map was generated based on transcriptome data (Figure 1a). TF open reading frames (ORFs) were amplified by RT-PCR and then electrophoresed on a 0.1% agarose gel.

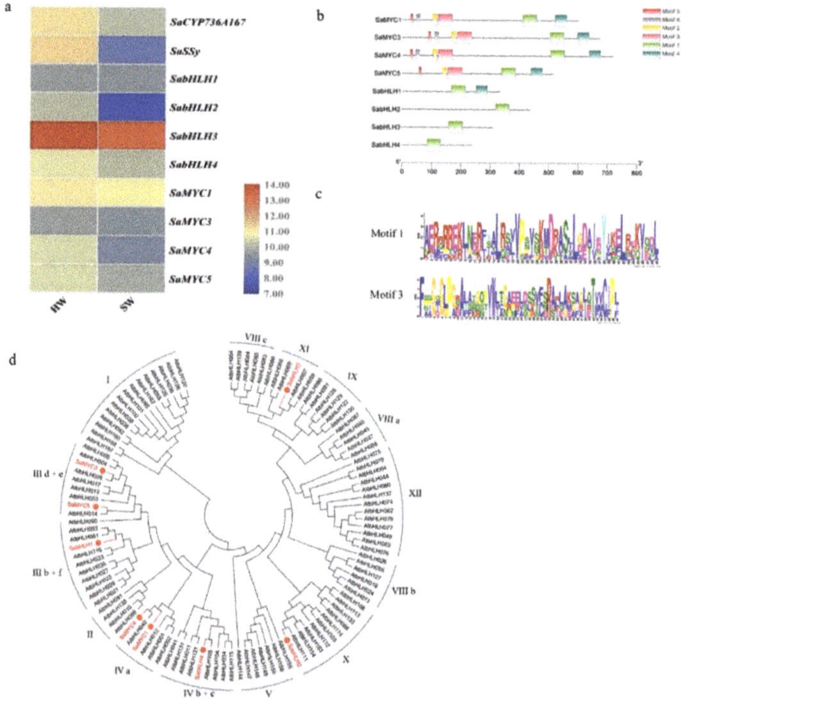

Figure 1. Sequence analysis of bHLH transcription factors and their co-expression patterns with *SaSSy* and *SaCYP736A167* in *Santalum album*. (**a**) Expression levels of bHLH transcription factors, *SaSSy* and *SaCYP736A167*. 10-year-old sandal tree was sampled. The heat map was generated based on log2-transformed count value from FPKM of transcriptome data using TBtools. HW: heartwood; SW: sapwood; (**b**) Schematic diagram of SabHLH motifs; (**c**) Motif 1 and Motif 3 domains; Motif 1 represents the bHLH domain, Motif 3 represents the MYC domain; (**d**) Phylogenetic analysis of SabHLH proteins in *Santalum album*. The bHLH transcription factors of *Santalum album* (marked with red) and *Arabidopsis thaliana* (black) were aligned by ClustalX 2.0, and the NJ (Neighbour-Joining) tree was constructed using MEGA 7.0 with 1000 bootstrap replicates.

3.2. Analysis of the Physicochemical Properties of SabHLH Proteins

The amino acid length of these eight TFs was 235–716 aa and the molecular weight (MW) was 25.87–77.72 kDa. Most of these proteins were acidic; the isoelectric point (pI) of seven proteins was less than 7.0, and the average hydrophilic value of these proteins was less than 0, indicating that they were all hydrophobic. The instability coefficient of these proteins was greater than 40, suggesting that they were unstable proteins (Table 1).

Table 1. Analysis of physicochemical properties of SabHLH proteins in *Santalum album*.

Gene Name	ORF Length (bp)	Amino Acid Length (aa)	Mw (kDa)	pI	Instability Index	Grand Average of Hydropathicity
SaMYC1	1800	599	67.32	5.69	52.17	−0.358
SaMYC3	2022	673	72.92	6.12	55.76	−0.533
SaMYC4	2151	716	77.72	5.12	64.21	−0.558
SaMYC5	1542	513	55.71	5.69	48.26	−0.404
SabHLH1	999	332	37.11	4.67	62.30	−0.456
SabHLH2	1302	433	47.54	6.09	46.59	−0.700
SabHLH3	927	308	32.09	5.91	51.81	−0.374
SabHLH4	708	235	25.87	7.71	51.73	−0.766

3.3. Conservative Motif Analysis of SabHLH Proteins

All eight TFs had highly conserved bHLH domain Motifs (Figure 1b). SaMYC1, SaMYC3, SaMYC4, and SaMYC5 also had highly conserved MYC domains (Figure 1c).

3.4. Phylogenetic Analysis of SabHLH Proteins

SaMYC1 and SaMYC4 were grouped with members of the *Arabidopsis* IVa subfamily, SaMYC3 and SaMYC5 were grouped with members of the *Arabidopsis* III d + e subfamily, SabHLH1 was grouped with members of the *Arabidopsis* III b + f subfamily, and SabHLH2 was grouped with members of the *Arabidopsis* X subfamily, while SabHLH3 was grouped with members of the *Arabidopsis* XI subfamily, and SabHLH4 was grouped with members of the *Arabidopsis* IV b + c subfamily (Figure 1d).

3.5. Subcellular Localization Analysis

A yellow fluorescent protein signal was observed under a laser confocal microscope. In the control group, fluorescence was expressed throughout the entire cell. The fused protein fluorescence signals of SaMYC3, SaMYC5, SabHLH1, SabHLH2, SabHLH3, SabHLH4 and YFP were detected in the nucleus, matching well with the fluorescence signal of mCherry protein located in the nucleus (Figure 2). This indicated that the six TFs were localized in the nucleus. However, SaMYC1 was localized in both the nucleus and the cytoplasm (Figure 2). Finally, the SaMYC4 YFP fluorescence signal was evenly distributed in the cytoplasm, indicating that it was localized in the cytoplasm (Figure 2).

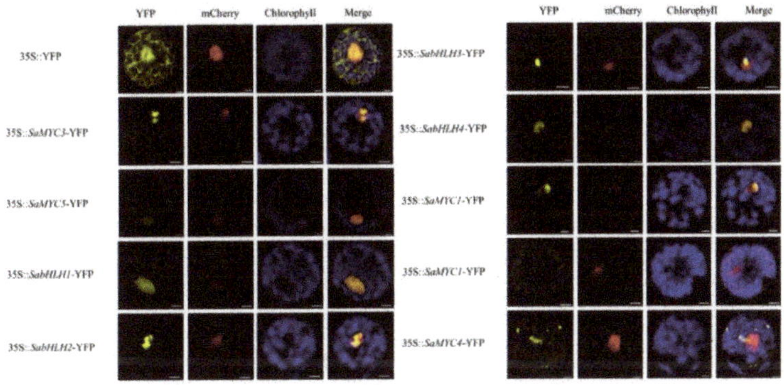

Figure 2. Subcellular location of SabHLHs in *Santalum album*. Note: 35S::YFP was the localization of unloaded vector, yellow fluorescence was YFP fluorescence (indicating protein localization), red fluorescence indicated nuclear-localized protein, blue fluorescence indicated chloroplast autofluorescence, orange with blue fluorescence was a merged image. Scale bars = 5 μm.

3.6. *SaMYC1* Conservative Domain Prediction

It was shown that *SaMYC1* had a typical bHLH-MYC_N superfamily domain at the position of 15–196, and a typical basic-helix-loop-helix domain at the position of 417–460 (Figure 3).

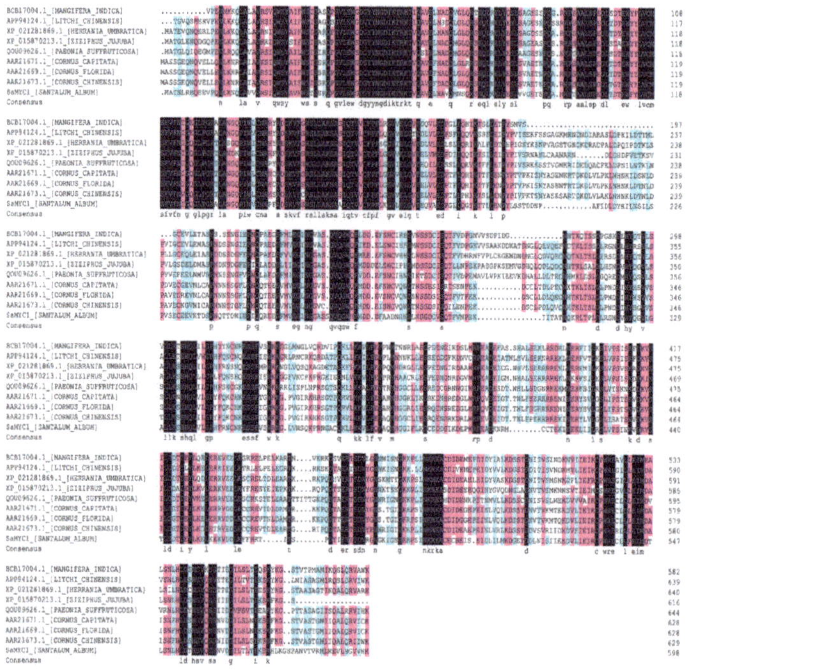

Figure 3. Similarity analysis of the amino-acid sequence of SaMYC1.

3.7. Analysis of G-Box Elements and E-Box Elements in Promoters *SaSSy* and *SaCYP736A167*

The *SaSSy* promoter region had no E-box element; it only contained two identical CAEs, the G-box (CACGTT), which could bind to the bHLH TF, and both were located on the antisense chain (Supplementary Figure S2a). There were six CAEs in the *SaCYP736A167* promoter region that could bind to the bHLH TF. There was no E-box element, but there were only three types of G-box elements, namely CACGTT, CACGTG, and CACGTA, two of which were on the antisense chain and four of which were on the sense chain (Supplementary Figure S2b).

3.8. Verification of the Activity of Transcription Factors

The results showed that the control group grew in a well growth status on the AbA-free SD/Leu plates; it was indicated that the yeast strain used in the experiment was intact and active. *SaMYC1* and *SabHLH1* were growing well on SD/-Leu/AbA 0 ng/mL plates and not on SD/-Leu/AbA 200 ng/mL plates, indicating that these two transcription factors were active in yeast and had no self-activation, which could be used for subsequent experiments. *SaMYC4*, *SaMYC5*, *SabHLH3* were not growing well in SD/-Leu/AbA 0 ng/mL plates and should not be used for subsequent yeast single-hybridization experiments. *SaMYC3*, *SabHLH2*, and *SabHLH4* were not growing well at all on SD/-Leu/AbA 0 ng/mL plates, indicating that they may not be active in yeast and cannot be used for subsequent experiments (Supplementary Figure S2).

3.9. Screening AbA Concentrations That Inhibiting the Growth of Bait Strains

The bait strains pAbAi-SSy1G-box, pAbAi-CYP1G- box, pAbAi-CYP2G-box, pAbAi-CYP3G-box, pAbAi-CYP4G-box were obtained in *E. coli*. The bait strains p53-AbAi, pAbAi-SSy1G-box, pAbAi-CYP1G-box, pAbAi-CYP2G-box, pAbAi-CYP3G-box, and pAbAi-CYP4G-box grew well in the presence or absence of AbA. However, the growth of pAbAi-SSy1G-box and pAbAi-CYP1G-box strains were completely inhibited, and the minimum AbA concentration to screen them was 100 ng/mL. pAbAi-CYP2G-box, pAbAi-CYP3G-box and pAbAi-CYP4G-box were all able to grow normally at 1000 ng/mL of AbA, indicating that endogenous TFs in yeast had been binded to these bait strains, so they could not be used for subsequent experiments. The growth of the positive control (p53-AbAi strain) was completely inhibited as AbA concentration was 200 ng/mL, indicating that was the minimum AbA screening concentration, so the yeast one-hybridization system was feasible (Supplementary Figure S3).

3.10. Assessment of the AbA Concentration That Inhibits the Growth of G-Box Mutant Element Bait Strains

The growth of the bait strain pAbAi-mSSy1G-box was completely inhibited as AbA concentration was 1000 ng/mL. This indicated that the screening concentration could be used in following experiments to preserve the mutant bait strain with a final concentration of 30% glycerol (Supplementary Figure S4).

3.11. Detection of the Interaction between SaMYC1 and Mutant G-Box Elements

The positive control p53-AbAi + pGADT7-Rec-53 could grow well on SD/-Leu medium supplemented with 200 ng/mL AbA. However, *SaMYC1* + pAbAi-mSSy1G-box and the negative control pGADT7 + pAbAi-mSSy1G-box could not grow on the screening medium with 1000 ng/mL AbA, *SaMYC1* + pAbAi-mCYP1G-box and the negative control pGADT7 + pAbAi-mCYP3G-box could not grow on the screening medium with 200 ng/mL AbA, while *SaMYC1* + pAbAi-mCYP4G-box and the negative control pGADT7 + pAbAi-mCYP4G-box could not grow on the screening medium supplemented with 500 ng/mL AbA (Figure 4c–e). This showed that *SaMYC1* could not bind to the mutant elements *mSSy1G*, *mCYP3G*, and *mCYP4G*, and it also illustratrd that *SaMYC1* could combine the G-box element in *SaSSy* and *SaCYP736A167* promoter (Figure 4b).

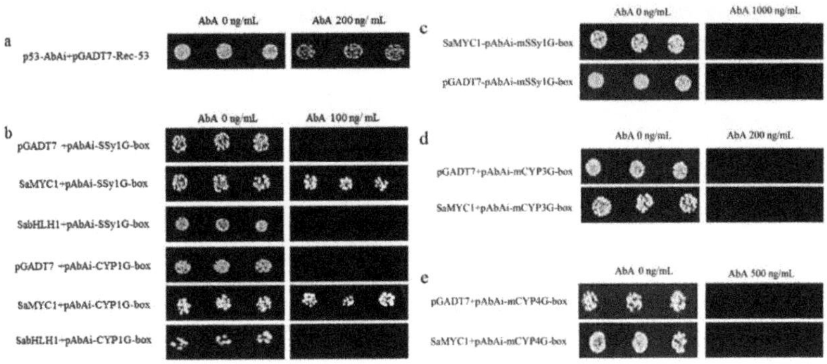

Figure 4. Interaction between SaMYC1 and G-box. Note: (**a**) Positive control; (**b**) Empty vector pGADT7 + pAbAi-SSy1G-box, pGADT7 + pAbAi-CYP1G-box was the negative control. (**c**) empty vector pGADT7 + pAbAi-mSSy1G-box was negative control; (**d**) empty vector pGADT7 + pAbAi-mCYP3G-box was negative control; (**e**) pGADT7 + pAbAi-mCYP3G-box was negative control; AbA concentrations at 0 ng/mL, 100 ng/mL, 200 ng/mL, 500 ng/mL, 1000 ng/mL, respectively, were the SD/-Leu screening media.

3.12. *SaMYC1 Activated SaSSy and SaCYP736A167 Promoter Activity*

As *SaMYC1* was bound to the *SaSSy* promoter, the ratio of *LUC/REN* was 1.85-fold higher than that of the control group, indicating that *SaMYC1* may bind to the *SaSSy* promoter and activate it (Figure 5b). When *SaMYC1* was bound to the *SaCYP736A167* promoter, the *LUC/REN* ratio was 1.55-fold higher than that of the control group, indicating that *SaMYC1* may also bind to the *SaCYP736A167* promoter and activate it (Figure 5a). These findings indicated that *SaMYC1* had activating effects on *SaSSy* and the *SaCYP736A167* promoters.

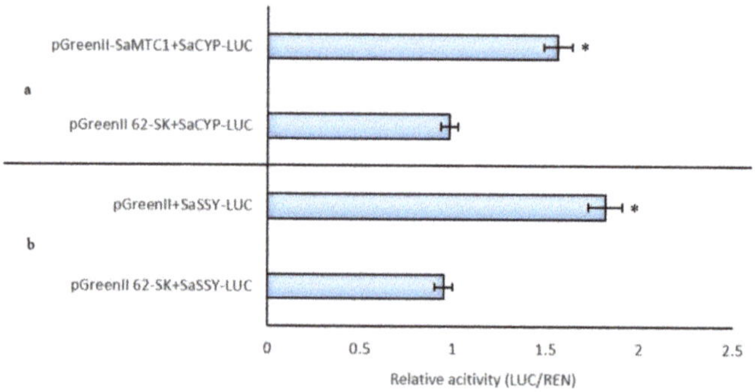

Figure 5. Dual-Luc test verified that *SaMYC1* activated the transcription of *SaSSy* and *SaCYP736A167*. Note: (**a**): Dual-Luc test verified that *SaMYC1* promoted the transcription of *SaCYP736A167*; (**b**): Dual-Luc test verified that *SaMYC1* promoted the transcription *SaSSy* in *Santalum album*. * indicated significant differences using *t*-test (* $p < 0.05$).

4. Discussion

bHLH TFs play integral roles in the resistance of a plant to environmental stress. *PebHLH35*, a bHLH TF gene, which localized in the nucleus of *Populus diversifolia*, was induced by drought stress and ABA, while overexpression of *PebHLH35* significantly improved drought tolerance [40]. Since overexpression of the bHLH TF gene *FtbHLH2* in *Fagopyrum tataricum* increased cold resistance in *A. thaliana*, it was suggested that this TF played a positive regulatory role in the resistance of *F. tataricum* to cold [41]. Overexpression of the *TabHLH39* gene in wheat significantly enhanced tolerance to salt stress in *A. thaliana* seedlings [42].

bHLH TFs were involved in regulating the synthesis of anthocyanins in *Triticum aestivum* [43]. In *A. thaliana* seedlings, bHLH TFs were involved in the anthocyanin biosynthetic pathway by forming TTG1/bHLH/MYB complexes with MYB TFs and WD40 proteins [44]. In *A. thaliana*, mutations in *MYC2*, *MYC3*, and *MYC4* downregulated gene expression involved in the regulation of glucosinolate biosynthesis [45]. In *Catharanthus roseus*, *CrMYC2* regulated the synthesis of alkaloids by the TF ORCA3, which contained AP2/ERF domains [46]. Overexpression of the *AabHLH1* gene localized in the nucleus of *Artemisia annua* upregulated the expression of structural genes, thereby increasing the accumulation of artemisinin [47]. In *A. thaliana*, *AtMYC2* was bound directly to the promoters of sesquiterpene synthetase genes *TPS21* and *TPS11*, activating their transcription. GA_3 and jasmonic acid signals could also integrate into the transcriptional regulation of sesquiterpene synthases, and regulate the synthesis of sesquiterpenes [48]. In *S. lycopersicum*, *SlMYC1* differentially regulated the biosynthesis of mono- and sesquiterpenes in the trichomes of leaves and stems, reversed regulation of sesquiterpene synthesis, and caused forward regulation of monoterpenoid synthesis [49]. In the woody plant *Aquilaria sinensis*, bHLH TFs were involved in regulating the synthesis of plant sesquiterpenes: one bHLH TF gene *AsMYC2* upregulated the expression of the sesquiterpene synthase gene *ASS1* in epidermal cells and the expression of *TPS21* and *TPS11* in *A. thaliana* [50].

bHLH TFs play a very important role in regulating the synthesis of plant sesquiterpenes, such as the synthesis of linalool in *Freesia hybrida* [51], linalool and β-caryophyllene in *Chimonanthus praecox* [52], and sesquiterpenes in *S. lycopersicum* [49]. In this study, based on the existing transcriptome data of our research group, the expression patterns of eight bHLH TFs were similar to those of two structural genes, *SaSSy* and *SaCYP736A167*, as screened by co-expression patterns. The eight TF genes were successfully cloned from a mixture of cDNA from the stems and leaves of *S. album*. This was used to construct a phylogenetic tree of these TFs together with members of the bHLH TF family in *A. thaliana*. The eight *S. album* TFs were mainly clustered into six subfamilies (Figure 1d), indicating that these TFs may perform different functions in sandal trees. TFs that were involved in the expression of regulatory structural genes generally bind to the promoter region of a structural gene, and this process typically took place in the nucleus. The subcellular localization results showed that SaMYC3, SaMYC5, SabHLH1, SabHLH2, SabHLH3, and SabHLH4 were all localized in the nucleus (Figure 2), consistent with previous reports in *F. hybrida* and *C. praecox* [51,52], indicating that they have typical characteristics of TFs. However, SaMYC1 was localized in both the nucleus and the cytoplasm while SaMYC4 was localized in the cytoplasm (Figure 2). The phylogenetic analysis indicated that SaMYC1, SaMYC4 and AtbHLH12 clustered in the IV a family (Figure 1d). Therefore, SaMYC1 and SaMYC4 may be able to modify the localization of other TFs such as AtMYC1, or the expression of genes that regulated structures in cells by binding to other TFs. However, this required further experimental verification.

At present, research on the molecular aspects of sandal oil biosynthesis has mainly focused on structural genes, with fewer studies on the transcriptional regulation of genes. In some plants, bHLH TFs played important roles in sesquiterpene biosynthesis [53]. To date, however, there were no reports of the involvement of bHLH TFs in the regulation of santalol biosynthesis in *S. album*. *SaSSy* and *SaCYP736A167* were two key genes-encoding enzymes that functioned downstream of the biosynthetic pathway of santalol sesquiterpenes. The sesquiterpene synthase encoded by *SaSSy* could ligate the substrate FPP into sesquiterpenes unique to santalol, and sandal sesquiterpenes could be oxidized to terpene under the action of oxidase encoded by the *SaCYP736A167* gene. Studies have shown that transcriptional regulation had amplification effects on structural gene functions. At present, in other plants, transcription factors regulated the synthesis of plant sesquiterpenes. For example, in cotton, *GaWRKY1* activated the CAD1-A promoter, thereby promoting the synthesis of sesquiterpenes [54]. In agarwood, the transcription factors *MYB4*, *WRKY4*, *MPKK2*, and *MAPK2* positively regulated the expression of the sesquiterpene synthase gene *ASS1-ASS3*, thereby promoting the synthesis of sesquiterpenes [50]. In *Artemisia annua*, overexpression of *AaWRKY1* activates the expression of the key enzyme gene *AaCYP71AV1*, thereby promoting the synthesis of artemisinin [55]. There were also many studies on the transcriptional regulation of bHLH transcription factors on the synthesis of sesquiterpenes, e.g., in *Artemisia annua*, overexpression of *AaMICC2* transcription factors could improve the transcription level of *AaCYP71AV1* and *DBR2* genes [56], and in Malus pumila calli, overexpression of *MdMYC2* and *MdERF3* could significantly increase the transcription levels of *MdHMGR2* and *MdAFS*, thereby increasing the synthesis of α-farneene [57]. However, it was unclear whether bHLH transcription factors in sandalwood were involved in transcriptional regulation of sandal sesquiterpene biosynthesis. Therefore, it is necessary to study the transcriptional regulation effect of bHLH transcription factors on *SaSSy* and *SaCYP736A167* promoters.

At this stage of this study, the effect of bHLH transcription factors on *SaSSy* and *SaCYP736A167* promoters was mainly explored by combining yeast one-hybridization experiments and dual luciferase experiments. In yeast one-hybridization experiments, *SaMYC1* was determined by screening the medium by transferring *SaMYC1* to two structural gene promoters by transferring *SaMYC1* into bait strains containing G-box elements SSy1G-box, CYP1G-box, and mutant G-box elements mSY1G-box, mCYP3G-box, and mCYP4G-box. The results showed that *SaMYC1* could be combined with the G-box components in *SaSSy* and *SaCYP736A167*. In the dual luciferase experiment, we have constructed the full length of the *SaSSy*, *SaCYP736A167* promoter and *SaMYC1* transcription factor into the reporter and effector carrier of the double luciferase, co-transformed the tobacco in an agrobacterium-mediated manner, and explored the effect of the transcription factor on the promoter of the two structural genes by detecting the chemiluminescence value of the protein of interest. The results showed that *SaMYC1* can activate *SaSSy* and *SaCYP736A167* promoters. Similar to *AtTT8* in *Arabidopsis* that could upregulate the structural genes *DFR* and *BAN* of flavonoid synthesis pathways to promote the biosynthesis of flavonoids in *Arabidopsis siliques* [58], *AtGL3* could upregulate the expression of *DFR*, a key structural gene for anthocyanin synthesis [59]. At the same time, many studies have shown that bHLH transcription factors often formed complexes with other transcription factors and were involved in the regulation of transcriptional expression of structural genes. In *Arabidopsis*, bHLH formed complexes with *MYB* and *WD40* subunits to regulate the expression of structural genes along the synthesis pathways of anthocyanins and flavonoids, as well as stamens development and seed formation [59–61].

In this study, yeast single-hybridization experiments were tested to find that *SaMYC1* could bind to the G-box element in the promoter of the santalol biosynthetic key enzyme gene *SaSSy*. Double luciferase experiments were used to show that *SaMYC1* could activate the *SaSSy* promoter, it was speculated that *SaMYC1* was a positive regulator of the key enzyme gene in santalol biosynthetic pathway. Results from *A. thaliana* [61], *Vitis vinifera* L. [62], and *Aquilaria sinensis* Lour. [50] indicated that bHLH transcription factors often co-regulated the expression of structural genes in conjunction with other transcription factors. Therefore, our study could also lay a foundation for subsequent exploration of whether *SaMYC1* regulated the expression of *SaSSy* with other transcription factors and how to co-regulate *SaSSy*.

5. Conclusions

bHLH TF genes with similar expression patterns and high expression levels were screened by co-expression analysis. All eight TFs had highly conserved bHLH domains and *SabHLH1*, *SabHLH2*, *SabHLH3*, and *SabHLH4*, had highly conserved MYC domains. It was indicated that the eight genes belonged to six subfamilies of the bHLH TF family. Among them, SaMYC1 was found in both the nucleus and the cytoplasm, while SaMYC4 was only localized in the cytoplasm. The remaining six TFs were localized in nucleus. *SaMYC1* could bind to the G-box of *SaSSy* and the *SaCYP736A167* promoter and the *LUC/REN* value was 1.85- or 1.55-fold higher, respectively, than that of the control group. It was inferred that *SaMYC1* could activate both *SaSSy* and *SaCYP736A167* promoters.

Supplementary Materials: The following supporting information can be downloaded at: https://www.mdpi.com/article/10.3390/life12071017/s1. Supplementary Table S1: NCBI login numbers for transcription factors in *Santalum album*; Supplementary Table S2: Premier information in *Santalum album*; Supplementary Table S3: The cluster multiple sequence alignment of SabHLHs in *Santalum album*; Supplementary Table S4: Yeast one-hybrid experimental genes synthesis sequences in *Santalum album*. Supplementary Figure S1: TF open reading frames (ORFs) were amplified by RT-PCR and then electrophoresed on a 0.1% agarose gel. Supplementary Figure S2: Detection of G-box elements of SaSSy ((a) G-box marked blue) and SaCYP736A167 ((b) G-box marked pink) promoter. Supplementary Figure S3: Transcription factor activity verification. Supplementary Figure S4: Screening of the inhibitory AbA concentrations of bait strain. Note: p53-AbAi was the positive control; pAbAi-SSy1G-box was the G-box element strain of SaSSy; pAbAi-CYP1G-box was the G-box element strain of SaCYP736A67; the SD/-Leu selection media with AbA concentration of 0 ng/mL, 50 ng/mL, 100 ng/mL, 200 ng/mL, respectively. Supplementary Figure S5: Screening AbA concentration of mutant G-box element bait strain. (a): Screening of the lowest inhibitory AbA concentration of mutant bait strain pAbAi-mSSy1G-box; (b): Screening of the lowest inhibitory AbA concentration of mutant bait strain CYP3G-box; (c): Screening of the lowest inhibitory AbA concentration of mutant bait strain pAbAi-mCYP3G-box. A-L: The SD/-Leu selection media with AbA concentration of 0 ng/mL,

50 ng/mL, 100 ng/mL, 200 ng/mL, 300 ng/mL, 400 ng/mL, 500 ng/mL, 600 ng/mL, 700 ng/mL, 800 ng/mL, 900 ng/mL, 1000 ng/mL, respectively.

Author Contributions: T.Z., X.C., H.Y., Y.Z., M.N., Y.X., X.Z., Y.L. and G.M. designed the experiment and provided guidance for the study. T.Z., H.Y. and X.C. prepared samples for all analyses. T.Z. and H.Y. conducted statistical analyses. T.Z. and G.M. co-wrote the manuscript. G.M. interpreted the data. G.M. assessed the experimental results. All authors have read and agreed to the published version of the manuscript.

Funding: This work was financially supported by Guangdong Key Areas Biosafety Project (2022B1111040003), National Key Research & Development Program of China (2021YFC3100400) and the National Natural Science Foundation of China (grant numbers 32171841, 32101512 and 32100311).

Institutional Review Board Statement: Not applicable.

Informed Consent Statement: Not applicable.

Data Availability Statement: All data generated or analyzed during this study are included in this published article.

Acknowledgments: The authors thank the assistance and advice provided by Jaime A. Teixeira da Silva (independent researcher, Japan) on earlier versions of the paper.

Conflicts of Interest: The authors declare no conflict of interest.

References

1. Jin, J.; Zhang, H.; Kong, L.; Gao, G.; Luo, J. PlantTFDB 3.0: A portal for the functional and evolutionary study of plant transcription factors. *Nucl. Acid. Res.* **2014**, *42*, 1182–1187. [CrossRef] [PubMed]
2. Chuang, Y.C.; Hung, Y.C.; Tsai, W.C.; Chen, W.H.; Chen, H.H. PbbHLH4 regulates floral monoterpene biosynthesis in *Phalaenopsis* orchids. *J. Exp. Bot.* **2018**, *69*, 4363–4377. [CrossRef] [PubMed]
3. Yan, H.F.; Li, M.Z.; Xiong, Y.P.; Wu, J.M.; Teixeira da Silva, J.A.; Ma, G.H. Genome-wide characterization, expression profile analysis of WRKY family genes in *Santalum album* and functional identification of their role in abiotic stress. *Int. J. Mol. Sci.* **2019**, *20*, 5676. [CrossRef] [PubMed]
4. Nieuwenhuizen, N.J.; Chen, X.; Wang, M.Y.; Matich, A.J.; Perez, R.L.; Allan, A.C.; Green, S.A.; Atkinson, R.G. Natural variation in monoterpene synthesis in kiwifruit: Transcriptional regulation of terpene synthases by nac and ethylene-insensitive3-like transcription factors. *Plant Physiol.* **2015**, *167*, 1243–1258. [CrossRef] [PubMed]
5. Jones, S. An overview of the basic helix-loop-helix proteins. *Genom. Biol.* **2004**, *5*, 266. [CrossRef] [PubMed]
6. Carretero-Paulet, L.; Galstyan, A.; Roig-Villanova, I.; Carretero-Paulet, L.; Galstyan, A.; Roig-Villanova, I.; Martínez-García, J.F.; Bilbao-Castro, J.R.; Robertson, D.L. Genome-wide classification and evolutionary analysis of the bHLH family of transcription factors in *Arabidopsis*, poplar, rice, moss, and algae. *Plant Physiol.* **2010**, *153*, 1398–1412. [CrossRef]
7. Wang, J.Y.; Hu, Z.Z.; Zhao, T.M.; Yang, Y.W.; Chen, T.Z.; Yang, M.L.; Yu, W.G.; Zhang, B.L. Genome-wide analysis of bHLH transcription factor and involvement in the infection by yellow leaf curl virus in tomato (*Solanum lycopersicum*). *BMC Genom.* **2015**, *16*, 39. [CrossRef] [PubMed]
8. Nathalie, N.; Debeaujon, I.; Jond, C.; Pelletier, G.; Caboche, M.; Lepiniec, L. The *TT8* gene encodes a basic helix-loop-helix domain protein required for expression of *DFR* and *BAN* genes in *Arabidopsis* siliques. *Plant Cell* **2014**, *12*, 1863–1878.
9. Zhu, Z.; Liang, H.; Chen, G.; Li, F.; Hu, Z. The bHLH transcription factor *SlPRE2* regulates tomato fruit development and modulates plant response to gibberellin. *Plant Cell Rep.* **2019**, *38*, 1053–1064. [CrossRef]
10. Qi, T.-C.; Huang, H.; Song, S.-S.; Xie, D. Regulation of jasmonate-mediated stamen development and seed production by a bHLH-MYB complex in *Arabidopsis*. *Plant Cell* **2015**, *27*, 1620–1633. [CrossRef]
11. Pillitteri, L.J.; Torii, K.U. Breaking the silence: Three bHLH proteins direct cell-fate decisions during stomatal development. *Bioessays* **2007**, *29*, 861–870. [CrossRef]
12. Chater, C.C.C.; Caine, R.S.; Fleming, A.J.; Gray, J.E. Origins and evolution of stomatal development. *Plant Physiol.* **2007**, *174*, 624–638. [CrossRef]
13. Schmitz, G.; Theres, K. Shoot and inflorescence branching. *Curr. Opin. Plant Biol.* **2005**, *8*, 506–511. [CrossRef]
14. Duek, P.D.; Fankhauser, C. bHLH class transcription factors take centre stage in phytochrome signalling. *Trends Plant Sci.* **2005**, *10*, 51–54. [CrossRef]
15. Friedrichsen, D.M.; Nemhauser, J.; Muramitsu, T.; Maloof, J.N.; Alonso, J.; Ecker, J.R.; Furuya, M.; Chory, J. Three redundant brassinosteroid early response genes encode putative bHLH transcription factors required for normal growth. *Genetics* **2002**, *162*, 1445–1456. [CrossRef]
16. Abe, H.; Urao, T.; Ito, T.; Seki, M.; Shinozaki, K.; Yamaguchi-Shinozaki, K. Arabidopsis AtMYC2 (bHLH) and AtMYB2 (MYB) function as transcriptional activators in abscisic acid signaling. *Plant Cell* **2003**, *15*, 63–78. [CrossRef]

17. Chini, A.; Fonseca, S.; Fernández, G.; Adie, B.; Chico, J.M.; Lorenzo, O.; García-Casado, G.; López-Vidriero, I.; Lozano, F.M.; Ponce, M.R.; et al. The JAZ family of repressors is the missing link in jasmonate signalling. *Nature* **2007**, *448*, 666–671. [CrossRef]
18. Hu, D.G.; Yu, J.Q.; Han, P.L.; Hu, D.-G.; Yu, J.Q.; Han, P.L.; Xie, X.B.; Sun, C.H.; Zhang, Q.Y.; Wang, J.H.; et al. The regulatory module MdPUB29-MdbHLH3 connects ethylene biosynthesis with fruit quality in apple. *N. Phytol.* **2019**, *221*, 1966–1982. [CrossRef]
19. Choi, H.; Oh, E. PIF4 Integrates multiple environmental and hormonal signals for plant growth regulation in *Arabidopsis*. *Mol. Cells* **2016**, *39*, 587–593. [CrossRef]
20. Lee, J.; Moon, S.; Jang, S.; Lee, J.; Moon, S.; Jang, S.; Lee, S.; An, G.; Jung, K.H.; Park, S.K. *OsbHLH073* negatively regulates internode elongation and plant height by modulating GA homeostasis in rice. *Plants* **2020**, *9*, 547. [CrossRef]
21. Kim, S.; Hwang, G.; Kim, S.; Thi, T.N.; Oh, E. The epidermis coordinates thermoresponsive growth through the phyB-PIF4-auxin pathway. *Nature Comm.* **2020**, *11*, 1053. [CrossRef]
22. Dozmorov, M.G.; Yang, Q.; Wu, W.J.; Wren, J.; Suhail, M.M.; Woolley, C.L.; Young, D.G.; Fung, K.M.; Lin, H.K. Differential effects of selective frankincense (Ru Xiang) essential oil versus non-selective sandalwood (Tan Xiang) essential oil on cultured bladder cancer cells: A microarray and bioinformatics study. *Chin. Med.* **2014**, *9*, 18. [CrossRef]
23. Santha, S.; Dwivedi, C. Anticancer effects of sandalwood (*Santalum album*). *Anticanc. Res.* **2015**, *35*, 3137–3145.
24. Scartezzini, P.; Speroni, E. Review on some plants of Indian traditional medicine with antioxidant activity. *J. Ethnoph.* **2000**, *71*, 23–43. [CrossRef]
25. Soneja, A.; Kaushik, P.; Kaushik, D.; Kumar, S.; Kumar, D. Antioxidant, Analgesic and Anti-inflammatory Activities of *Santalum album* Linn. *Plant. Med.* **2009**, *75*, 452–453. [CrossRef]
26. Suganya, K.; Liu, Q.F.; Koo, B.S. *Santalum album* extract exhibits neuroprotective effect against the TLR3-mediated neuroinflammatory response in human SH-SY5Y neuroblastoma cells. *Phytother. Res.* **2021**, *35*, 1991–2004. [CrossRef]
27. Han, X.; Beaumont, C.; Stevens, N. Chemical composition analysis and in vitro biological activities of ten essential oils in human skin cells. *Biochim. Open* **2017**, *5*, 1–7. [CrossRef]
28. Moy, R.L.; Levenson, C. Sandalwood album oil as a botanical therapeutic in dermatology. *J. Clin. Aesth. Derm.* **2017**, *10*, 34–39.
29. Baldovini, N.; Delasalle, C.; Joulain, D. Phytochemistry of the heartwood from fragrant *Santalum* species: A review. *Flav. Frag. J.* **2011**, *26*, 7–26. [CrossRef]
30. Jones, C.G.; Moniodis, J.; Zulak, K.G.; Scaffidi, A.; Plummer, J.A.; Ghisalberti, E.L.; Barbour, E.L.; Bohlmann, J. Sandalwood fragrance biosynthesis involves sesquiterpene synthases of both the terpene synthase (TPS)-a and TPS-b subfamilies, including santalene synthases. *J. Biol. Chem.* **2011**, *286*, 17445–17454. [CrossRef] [PubMed]
31. Diaz-Chavez, M.L.; Moniodis, J.; Madilao, L.L.; Jancsik, S.; Keeling, C.I.; Barbour, E.L.; Ghisalberti, E.L.; Plummer, J.A.; Jones, C.G.; Bohlmann, J. Biosynthesis of sandalwood oil: *Santalum album CYP76F* Cytochromes P450 produce santalols and bergamotol. *PLoS ONE* **2013**, *8*, e75053. [CrossRef]
32. Celedon, J.M.; Bohlmann, J. Genomics-based discovery of plant genes for synthetic biology of terpenoid fragrances: A case study in sandalwood oil biosynthesis. In *Synthetic Biology and Metabolic Engineering in Plants and Microbes*; Academic Press: Cambridge, MA, USA, 2016; Volume 576, pp. 47–67.
33. Srivastava, P.L.; Daramwar, P.P.; Krithika, R.; Pandreka, A.; Shankar, S.S.; Thulasiram, H.V. Functional characterization of novel sesquiterpene synthases from Indian sandalwood, *Santalum album*. *Sci. Rep.* **2015**, *5*, 10095. [CrossRef]
34. Zhang, X.H.; Niu, M.Y.; Teixeira da Silva, J.A.; Zhang, Y.Y.; Yuan, Y.F.; Jia, Y.X.; Xiao, Y.Y.; Li, Y.; Fang, L.; Zeng, S.J.; et al. Identification and functional characterization of three new terpene synthase genes involved in chemical defense and abiotic stresses in *Santalum album*. *BMC Plant Biol.* **2019**, *19*, 115. [CrossRef]
35. Zhang, X.H.; Teixeira da Silva, J.A.; Niu, M.Y.; Zhang, T.; Liu, H.F.; Zheng, F.; Yuan, Y.F.; Li, Y.; Fang, L.; Zeng, S.J.; et al. Functional characterization of an Indian sandalwood (*Santalum album* L.) dual-localized bifunctional nerolidol/linalool synthase gene involved in stress response. *Phytochemistry* **2021**, *183*, 112610. [CrossRef]
36. Niu, M.Y.; Yan, H.F.; Xiong, Y.P.; Zhang, Y.Y.; Zhang, X.H.; Li, Y.; Teixeira da Silva, J.A.; Ma, G.H. Cloning, characterization, and functional analysis of acetyl-CoA C-acetyltransferase and 3-hydroxy-3-methylglutaryl-CoA synthase genes in *Santalum album*. *Sci. Rep.* **2021**, *11*, 1082. [CrossRef]
37. Niu, M.Y.; Xiong, Y.P.; Yan, H.F.; Zhang, X.H.; Li, Y.; Teixeira da Silva, J.A.; Ma, G.H. Cloning and expression analysis of mevalonate kinase and phosphomevalonate kinase genes associated with the MVA pathway in *Santalum album*. *Sci. Rep.* **2021**, *11*, 16913. [CrossRef]
38. Zhang, Y.Y.; Yan, H.F.; Li, Y.; Xiong, Y.P.; Niu, M.Y.; Zhang, X.H.; Teixeira da Silva, J.A.; Ma, G.H. Molecular cloning and functional identification of 1-deoxy-D-xylulose 1,5-phosphate reductoisomerase from *Santalum album*. *Genes* **2021**, *12*, 626. [CrossRef]
39. Chen, Y.; Jiang, T.; Zhang, H.; Fu, S.L.; Zhao, G.Y.; Jia, J.Z.; Kong, X.Y. LRRC31 inhibits DNA repair and sensitizes breast cancer brain metastasis to radiation therapy. *Nat. Cell Biol.* **2020**, *22*, 1276–1285. [CrossRef]
40. Dong, Y.; Wang, C.; Han, X.; Han, X.; Tang, S.; Liu, S.; Xia, X.L.; Yin, W.L. A novel bHLH transcription factor *PebHLH35* from *Populus euphratica* confers drought tolerance through regulating stomatal development, photosynthesis and growth in *Arabidopsis*. *Biochem. Biophys. Res. Comm.* **2014**, *450*, 453–458. [CrossRef]
41. Yao, P.F.; Sun, Z.X.; Li, C.L.; Li, M.; Deng, R.; Huang, Y.; Zhao, H.; Chen, H.; Wu, Q. Overexpression of *Fagopyrum tataricum FtbHLH2* enhances tolerance to cold stress in transgenic *Arabidopsis*. *Plant Physiol. Biochem.* **2018**, *125*, 85–94. [CrossRef]
42. Zhai, Y.; Zhang, L.; Xia, C.; Chuan, S.L.; Zhao, G.Y.; Kong, X.Y. The wheat transcription factor, TabHLH39, improves tolerance to multiple abiotic stressors in transgenic plants. *Biochem. Bioph. Res. Comm.* **2016**, *473*, 1321–1327. [CrossRef]

43. Ludwig, S.R.; Habera, L.F.; Dellaporta, S.L.; Wessler, S.R. *Lc*, a member of the maize *R* gene family responsible for tissue-specific anthocyanin production, encodes a protein similar to transcriptional activators and contains the *myc*-homology region. *Proc. Nat. Acad. Sci. USA* **1989**, *86*, 7092–7096. [CrossRef]
44. Gonzalez, A.; Zhao, M.; Leavitt, J.M.; Lloyd, A.M. Regulation of the anthocyanin biosynthetic pathway by the TTG1/bHLH/Myb transcriptional complex in *Arabidopsis* seedlings. *Plant J.* **2008**, *53*, 814–827. [CrossRef]
45. Schweizer, F.; Fernández-Calvo, P.; Zander, M.; Diez-Diaz, M.; Fonseca, S.; Glauser, G.; Lewsey, M.G.; Ecker, J.R.; Solano, R.; Reymond, P. *Arabidopsis* basic helix-loop-helix transcription factors MYC2, MYC3, and MYC4 regulate glucosinolate biosynthesis, insect performance, and feeding behavior. *Plant Cell* **2013**, *25*, 3117–3132. [CrossRef]
46. Zhang, H.T.; Hedhili, S.; Montiel, G.; Zhang, Y.X.; Guillaume, C.; Martial, P.; Pascal, G.; Memelink, J. The basic helix-loop-helix transcription factor CrMYC2 controls the jasmonate-responsive expression of the *ORCA* genes that regulate alkaloid biosynthesis in *Catharanthus roseus*. *Plant J.* **2011**, *67*, 61–71. [CrossRef]
47. Ji, Y.; Xiao, J.; Shen, Y.; Ma, D.M.; Li, Z.Q.; Pu, G.B.; Li, X.; Huang, L.L.; Liu, B.Y.; Ye, H.C.; et al. Cloning and characterization of AabHLH1, a bHLH transcription factor that positively regulates artemisinin biosynthesis in *Artemisia annua*. *Plant Cell Physiol.* **2014**, *55*, 1592–1604. [CrossRef]
48. Hong, G.J.; Xue, X.Y.; Mao, Y.B.; Wang, L.J.; Chen, X.-Y. *Arabidopsis* MYC2 interacts with DELLA proteins in regulating sesquiterpene synthase gene expression. *Plant Cell* **2012**, *24*, 2635–2648. [CrossRef]
49. Xu, J.S.; van Herwijnen, Z.O.; Dräger, D.B.; Dräger, D.B.; Sui, C.; Haring, M.A.; Schuurink, R.C. SlMYC1 regulates type VI glandular trichome formation and terpene biosynthesis in tomato glandular cells. *Plant Cell* **2018**, *30*, 2988–3005. [CrossRef]
50. Xu, Y.H.; Liao, Y.C.; Lv, F.F.; Zhang, Z.; Sun, P.W.; Gao, Z.H.; Hu, K.P.; Sui, C.; Jin, Y.; Wei, J.H. Transcription factor AsMYC2 controls the jasmonate-responsive expression of *ASS1* regulating sesquiterpene biosynthesis in *Aquilaria sinensis* (Lour.) Gilg. *Plant Cell Physiol.* **2017**, *58*, 1924–1933. [CrossRef]
51. Yang, Z.; Li, Y.; Gao, F.; Jin, W.; Li, S.; Shadrack, K.; Yang, S.; Bao, T.; Gao, X.; Wang, L. MYB21 interacts with MYC2 to control the expression of terpene synthase genes in flowers of *Freesia hybrida* and *Arabidopsis thaliana*. *J. Exp. Bot.* **2020**, *71*, 4140–4158. [CrossRef] [PubMed]
52. Aslam, M.Z.; Lin, X.; Li, X.; Yang, N.; Chen, L. Molecular cloning and functional characterization of *CpMYC2* and *CpBHLH13* transcription factors from wintersweet (*Chimonanthus praecox* L.). *Plants* **2020**, *9*, 785. [CrossRef] [PubMed]
53. Hong, Y.Q.; Ahmad, N.; Tian, Y.Y.; Liu, J.; Li, H. Genome-wide identification, expression analysis, and subcellular localization of *Carthamus tinctorius* bHLH transcription factors. *Int. J. Mol. Sci.* **2019**, *20*, 3044. [CrossRef] [PubMed]
54. Xu, Y.H.; Wang, J.W.; Wang, S.; Wang, J.Y.; Chen, X.Y. Characterization of GaWRKY1, a cotton transcription factor that regulates the sesquiterpene synthase gene (+)-δ-cadinene synthase-A. *Plant Physiol.* **2004**, *135*, 507–515. [CrossRef]
55. Han, J.; Wang, H.; Lundgren, A.; Brodelius, P.E. Effects of overexpression of AaWRKY1 on artemisinin biosynthesis in transgenic *Artemisia annua* plants. *Phytochemistry* **2014**, *102*, 89–96. [CrossRef]
56. Tang, K.X.; Pan, Q.F.; Zhang, F.Y.; Fu, X.; Lv, Z.; Zhang, F.; Pan, Q.; Wang, G.; Sun, X.; Tang, K. The jasmonate-responsive AaMYC2 transcription factor positively regulates artemisinin biosynthesis in *Artemisia annua*. *N. Phytol.* **2016**, *210*, 1269–1281.
57. Wang, Q.; Liu, H.; Zhang, M.; Liu, S.H.; Zhang, Y.H. MdMYC2 and MdERF3 positively co-regulate α-Farnesene biosynthesis in Apple. *Front. Plant Sci.* **2020**, *11*, 512844. [CrossRef]
58. Feyissa, D.N.; Løvdal, T.; Olsen, K.M.; Slimestad, R.; Lillo, C. The endogenous GL3, but not EGL3, gene is necessary for anthocyanin accumulation as induced by nitrogen depletion in *Arabidopsis rosette* stage leaves. *Planta* **2009**, *230*, 747–754. [CrossRef]
59. Xie, Y.; Tan, H.; Ma, Z.; Huang, J.R. DELLA proteins promote anthocyanin biosynthesis via sequestering MYBL2 and JAZ suppressors of the MYB/bHLH/WD40 complex in *Arabidopsis thaliana*. *Mol. Plant* **2016**, *9*, 711–721. [CrossRef]
60. Xu, W.J.; Dubos, C.; Lepiniec, L. Transcriptional control of flavonoid biosynthesis by MYB-bHLH-WDR complexes. *Trends Plant Sci.* **2015**, *20*, 176–185. [CrossRef]
61. Baudry, A.; Heim, M.A.; Dubreucq, B.; Caboche, M.; Weisshaar, B.; Lepiniec, L. TT2, TT8, and TTG1 synergistically specify the expression of BANYULS and proanthocyanidin biosynthesis in *Arabidopsis thaliana*. *Plant J.* **2004**, *39*, 366–380. [CrossRef]
62. Hichri, I.; Heppel, S.C.; Pillet, J.; Léon, C.; Czemmel, S.; Delrot, S.; Lauvergeat, V.; Bogs, J. The basic helix-loop-helix transcription factor MYC1 is involved in the regulation of the flavonoid biosynthesis pathway in grapevine. *Mol. Plant* **2010**, *3*, 509–523. [CrossRef]

MDPI
St. Alban-Anlage 66
4052 Basel
Switzerland
www.mdpi.com

Life Editorial Office
E-mail: life@mdpi.com
www.mdpi.com/journal/life